300586605X

Appropriate Biotechnology in Small–scale Agriculture

How to reorient research and development

Appropriate Biotechnology in

Small-scale Agriculture

How to reorient research and development

Edited by

Joske F.G. Bunders and Jacqueline E.W. Broerse

Department of Biology and Society
Faculty of Biology
Free University
De Boelelaan
1081 HV Amsterdam
The Netherlands

C·A·B International

C·A·B International
Wallingford
Oxon OX10 8DE
UK

Tel: Wallingford (0491) 32111
Telex: 847964 (COMAGG G)
Telecom Gold/Dialcom: 84: CAU001
Fax: (0491) 33508

A catalogue entry for this book is available from the British Library

ISBN 0 85198 770 2

Printed and bound in the UK by Redwood Press Ltd, Melksham

Contents

Preface

This book is about small-scale farmers and the possibilities of biotechnology to improve their living conditions. Small-scale farmers (male and female) in developing countries are ingenious - but vulnerable. With little capital, limited access to external resources and only the most marginal of land, they endeavour to support their family's current needs. Stifled by the most unrewarding of conditions, they often live on the edge of survival. Technology could be a crucial factor in any improvement of their living conditions, but only if the technology is appropriate; successful technologies will both take into account the prevailing conditions and make the best use of local resources.

Biotechnology contains elements of an appropriate technology, but these remain largely unexplored. As practised now, biotechnology is oriented to the developed world. From scientific discovery onwards, biotechnology leads towards the profitable markets of high technology health care and intensive agriculture. Biotechnological applications currently in development will bring innovations mainly to the group of capital-rich, often large-scale producers. The focus on their farming systems implies that an increasing gap in yield and income between poor and rich farmers can be expected. This increased gap will make it more difficult for poor farmers to remain or become integrated in the local or international economies. In most developing countries these agricultural developments are not accompanied by successful creation of employment. An accelerated 'drop-out race' will, therefore, affect the availability of basic requirements of a large part of the population. The number of 'drop-outs' is not only dependent on the number of small-scale farmers, the agro-ecological conditions but also on infrastructure, agricultural policy measures and socio-economic conditions. Thus, in practice, biotechnology seems destined to have a largely negative impact on many of the small-scale farmers in developing countries.

The question is, whether it is possible to undertake activities that reduce the number of farmers that will not be able to catch up with agricultural developments. Can technologies be developed that improve their living conditions? Or more in general, is it possible to design technologies that are specifically appropriate for the resource-poor farmers? If yes, how can this be done?

Between the promise of biotechnology for small-scale farmers and its reality is a chasm of ignorance, misunderstanding and perceived conflict of interest. In the rare instances that research or projects are aimed at small-scale farmers, they are usually designed *for* them, not *with* them. Their lack of political and economic power precludes small-scale farmers from being able to influence the design and implementation of projects, or from redefining the direction of biotechnological research. If small-scale farmers in developing countries are to derive any benefit from biotechnology, some substantial changes are needed in the way technology development is funded and undertaken.

The first anomaly in current practice is that projects aimed at small-scale farmers but designed along conventional lines often fail because attention is not specifically paid to their small-scale farming systems, systems which are extremely complex and risk-prone. A key question for donor organizations and governments, therefore, should be: 'Is it possible to assess project proposals in a way which will increase the likelihood that projects undertaken in research facilities far distant from the farmer will bring real benefits for small-scale agriculture in developing countries?' And, if so, how does the design of that assessment need to depart from conventional practice? These questions are the subject of Part One of this book.

The second anomaly is that small-scale farmers are not a target group of mainstream (market-oriented) research. Emphasis is never placed in policy and research agendas on the generation of appropriate biotechnological innovations for small-scale farmers. It often is assumed that mainstream agricultural research benefits all farmers, including resource-poor small-scale farmers. In general, scientists and policy-makers are ignorant of the problems and needs of small-scale farmers and there is virtually no exchange of information between small-scale farmers, scientists, and policy-makers. The second key question therefore is: 'How can we enhance information exchange with the aim of incorporating the priorities of small-scale agriculture in biotechnology research policies of governments, research institutions, donor organizations, and other involved parties?' This question is the subject of Part Two of this book.

We hope this book will be read and used by national parliaments in developed countries, donor organizations, governments in developing countries, non-governmental organizations (NGOs), and institutions which intend to develop biotechnology projects for small-scale agriculture. We also seek to interest the industrial sector, which has a significant impact on biotechnological research and development in the Third World.

Acknowledgements

This publication was financially supported by the Netherlands Organization for Technology Assessment (NOTA). Lydi Sterrenberg of NOTA provided important support, guidance and patience for which we are very grateful.

Most helpful comments and suggestions for improvement of the text were provided by Gordon Banta, Dick Bijloo, Peter Hall, Bertus Haverkort, Doug Horton, Samuel Kassapu, Paul Leeflang, Vijay Paranjpye and Andrew Spurling.

We have especially benefited from comments by our colleague Theo van de Sande. We are also grateful to John Hodgson who undertook the English editing. Their valuable contribution is warmly acknowledged.

Chapter 1

The potential of biotechnology for small-scale agriculture

Jacqueline E.W. Broerse and Joske F.G. Bunders

1.1   Problems of small-scale farmers

Two-thirds of the Third World population depends on agriculture for its livelihood. Seventy-five percent of all farmers in developing countries are small-scale farmers. For a variety of reasons, some historical and others political, most small-scale farmers live in those parts of developing countries where the climatic, agro-ecological, biological, infrastructural and institutional conditions are unfavourable for agricultural production (Bunders *et al.*, 1990; World Bank, 1990).

Unfavourable climatic conditions

With few exceptions, small-scale farmers in developing countries live in areas of low and irregular rainfall, short rainy seasons, midsummer droughts, high intensity of rainfall and/or high temperature and consequent high evaporation. These conditions cause low yields, unreliable yields and complete crop failures. In livestock, these may cause low productivity and reproductivity.

Poor soil conditions

Most small-scale producers farm very poor soils. The limited amount of nutrients in these soils is often even more reduced by crop and livestock production which is too infrequently accompanied by the necessary replenishment of nutrients. The continuous soil nutrient depletion that takes place amounts to the progressive destruction of the most important resource basis of small-scale farming systems. Small-scale farmers are confronted not only with soil of diminished quality but indeed with the wholesale reduction in soil quantity through erosion. Exposed (top)soils are removed in huge amounts by rain and wind. Their exposure initially is closely related to the factors which exacerbate soil quality - the over-exploitation of arable land, overgrazing, extension of rainfed agriculture into dry-land areas, and deforestation. Soil erosion not only depletes soil fertility but also increases salinization of surface water and the likelihood of flooding. Furthermore, the absolute decrease of arable and grazing land reduces agricultural and livestock outputs.

Weeds, pests and diseases of crops and animals

The widespread prevalence of weeds, pests and diseases is seriously affecting crop and livestock production. Unlike their large-scale, commercial counterparts, small-scale farmers rarely use capital-intensive pesticides, herbicides, vaccines and medicines. If they do, it is usually only after much damage has already been caused to their farms. Furthermore, the farming systems of small-scale farmers are rendered more vulnerable to weeds, pests and diseases by practices forced upon them by circumstances: shortened fallow periods as a result of pressure on the land, poor rotation, continuous cropping without fertilizer, limited storage capacity, and the absence of irrigation, draught power and labour.

Adverse internal conditions

Malnutrition, illiteracy and shortage of draught power, equipment and labour all contribute to a low level of farm efficiency. Malnutrition, as a consequence of low crop and livestock production, heavy workload, and lack of money to buy additional food, is widespread among small-scale farmer families. Illiteracy is common among small-scale farmers in developing countries and can constrain any improvement to traditional agriculture. Shortages in draught power, equipment and labour lead to delays in cropping, thereby considerably reducing already-moderate yields.

Limited access to external inputs

Small-scale farmers usually live in remote parts of developing countries where infrastructure is relatively poor and external inputs are both difficult to obtain and expensive. Access to fertilizers, quality seeds, chemicals and tools, and to credits and transport is limited. Since small-scale farmers usually produce only a small surplus for the market, they cannot earn enough money to buy capital-intensive inputs. In any case, on small-scale farms, capital-intensive inputs often give only relatively low economic returns.

In addition, extension services often operate poorly. These services are sparsely staffed and cannot reach the whole small-scale farmer population; women, in particular, are neglected. In most cases, the agricultural practices disseminated by extension services do not suit small-scale farmers.

Limited access to markets

Small-scale farmers, as a disparate and uncoordinated group have a weak commercial bargaining position and continue to lose market share to better-equipped, better-organized larger scale producers. Small-scale farmers are considerably disadvantaged both in production and in marketing their products by a number of factors. Firstly, small-scale farmers lack knowledge on pricing. Secondly, there is a lack of transport facilities and, thirdly, there is a lack of local processing facilities that would enhance the quality and/or preservation.

Overviewing the situation, we may conclude that the farming systems of small-scale farmers are complex, the environments in which they work are diverse, and production is risk-prone. A relatively high population growth combined with increasing per capita use of the available natural resources (land, water and wood) have conspired to create intense pressure on the land and to devalue the living standards of those who rely on it. Despite the ingenuity of small-scale farmers and decades of development planning, the incidence of rural poverty in Third World countries is virtually unchanged: between 30 and 40 percent of the rural population still exists on incomes which do not ensure a minimum calorific intake (Bukman, 1989).

1.2   Assessment of biotechnology

Biotechnology can be defined as the integrated use of molecular genetics, biochemistry, microbiology and process technology employing micro-organisms, parts of micro-organisms, or cells and tissues of higher organisms to supply goods and services (DGIS, 1989). Biotechnology is neither a scientific discipline nor an industry but a continuum of technologies. It would appear that biotechnology could make not only a significant contribution to the solution of several problems of small-scale agriculture in developing countries but also it could make that contribution in an appropriate way (Bunders *et al.*, 1990). In contrast to biotechnological research and development (R&D) itself, many of its applications are inexpensive, uncomplicated and do not require capital- and energy-intensive inputs. Biotechnology is often flexible in scale and in the type of technology used, facilitating small-scale decentralized application and adaptation to the special circumstances of small-scale farmers (Broerse, 1990). Moreover, biotechnology could be linked to indigenous knowledge, existing practices and local initiatives, given the intrinsically biological nature of small-scale agriculture. Given their large numbers, any innovation which could be adopted even by a minority of small-scale farmers would have a significant impact on poverty. The neglect of the past means that whatever potential there is for increased production on small-scale farms, is still almost completely unexploited (Chambers and Ghildyal, 1985; Farrington and Martin, 1988).

Without looking in detail at the local situation of small-scale farmers in developing countries - their skills, resources, environment and political climate - it is not possible to predict in advance which of the range of biotechnological developments are likely to be of benefit to them. However, some general comments can be made. With the element of sustainability very much in mind, we have concentrated on those aspects of biotechnology which could be supported by research activities within developing countries. The major techniques of agricultural biotechnology described here are presented in order of increasing technical complexity. They include fermentation, microbial inoculation of plants, plant cell and tissue culture, enzyme technology, embryo transfer, protoplast fusion, hybridoma or monoclonal antibody technology, and rDNA technologies. For a technical overview of the technologies, we refer to the Appendix.

Fermentation

Fermentation (the production of substances with the aid of micro-organisms) is the most mature area of biotechnology. Fermentation is a widely-practised traditional means of food and beverage production in both developed and developing countries. In much of the broad range of fermentation processes employed, local raw materials are used.

In developing countries, there are many examples of domestic, small- and medium-scale industries applying techniques which have been available, in some cases, for centuries. Many developing countries have an extensive beer and wine industry (alcoholic fermentation). Solid-state fermentation, such as the production of tempeh (compact cake made from soyabean, groundnut or coconut), is widely practised in developing countries (Sasson, 1989). With some exceptions in East Asia, it operates at the village rather than at the agro-industrial level. Biogas digesters in their millions are in service in India and China (Senez, 1987). Biogas (an energy carrier) can replace firewood, contributing to efforts to combat deforestation and desertification.

There are also a number of examples of more recent developments in fermentation technology in developing countries. In Brazil and India, for example, there is large-scale production of ethanol as a liquid fuel (ethanol is produced by certain micro-organisms when they convert sugars in the absence of oxygen). The motivation, however, is political rather than economic; Brazil's bio-ethanol programme was intended to reduce imports of petrol and to provide an outlet for the sugar industry which was badly hit by the fall in the price of sugar on the world market (Primrose, 1987; Senez, 1987). In Cuba, single cell protein (microbial biomass) has been produced from agricultural raw materials. Single cell protein (SCP) would also be a useful food or food additive in other developing countries, where traditional food products are low in protein, or where the land is too arid to produce sufficient food of any type (Senez, 1987). In countries such as Thailand, amino acids (lysine) and antibiotics (kanamycin) have been produced (Yuthavong and Bhumiratana, 1989).

Medium- and large-scale production demands a considerable amount of equipment - including a fermentor, sterilization equipment, sensors, instrumentation and computers. The production of SCP, amino acids and secondary metabolites (compounds whose synthesis occurs after microbial growth ceases, e.g. antibiotics) is complicated; the initial capital investments are quite high and, once established, high-quality technical expertise and support facilities are required to maintain production

(Primrose, 1987). However, the facilities for alcoholic fermentation, solid-state fermentation and biogas production are neither sophisticated nor expensive. Traditional fermentation needs no control equipment. Instead, it can be conducted with the involvement of craftsmen or technicians who can judge through experience the state of fermentation.

Microbial inoculation of plants

Microbial inoculation involves the selection and multiplication of plant-beneficial micro-organisms and applying them to plants, seed or soil. The main uses of micro-organisms are as biofertilizers for improved plant nutrition and as biological control agents to combat pests, weeds and diseases. The prospects for improving agriculture through the use of microbial inocula are very good. With the possibility of better yields, lower costs and reduced dependence on chemicals, microbial inoculation of plants is likely to be of great importance, particularly in less-intensive, low-input agricultural systems in developing countries (Davison, 1988).

Biological nitrogen fixation

Biological nitrogen fixation (a process in which atmospheric nitrogen is fixed in organic compound by certain micro-organisms) holds much promise for developing countries and, indeed, is already being applied. For example, *Rhizobium* inoculants are produced for legumes in many developing countries. Inoculants for soyabean (grown for fodder, oil and export) account for a very large proportion of biological nitrogen fixation in developing countries. In some countries, inoculants for forage, pasture and food legumes are also produced. The extent to which externally-applied inoculants are an accepted farm practice varies widely. In any case, inoculant use currently has its main impact on large-scale producers (Eaglesham, 1989). Inoculants have had little effect on legume production where yields are poorest - at the small-scale farm level in developing countries. When, as is often the case, nitrogen is not the yield-limiting factor, it is difficult to assess the potential of nitrogen fixation. Nevertheless, *Rhizobium* inoculation is considered a key component in the improvement of pasture legume production in South America, West Asia and North Africa and it may prove beneficial elsewhere (Eaglesham, 1989).

The blue-green algae (cyanobacteria) are another source of biological nitrogen. They are distributed worldwide and contribute to the fertility of many agricultural ecosystems, either as free-living organisms or in symbiosis with the water fern *Azolla*. China and Vietnam have a long history of *Azolla* cultivation. The past decade, *Azolla* was introduced in paddy fields elsewhere in Asia including India, the Philippines and Thailand. The systems of *Azolla* application are diverse but always labour intensive. *Azolla* has an excellent potential for successful cultivation in irrigated deserts where humidity is relatively low; e.g. certain parts of North Africa and the Middle East. The utilization of inocula of free-living, blue-green algae, called algalization, is relatively easy but still limited. Research is being conducted in several Asian countries including China, the Philippines and India (Whitton and Roger, 1989).

Mycorrhizal associations

Mycorrhizal associations (symbioses between certain fungi and the roots of vascular plants) can increase the rate of uptake of nutrients such as phosphorus and nitrogen from deficient soils. The production of mycorrhizal inocula may well be possible in developing countries but it is not a common practice. There is some research on mycorrhiza in several developing countries, including Thailand (Yuthavong and Bhumiratana, 1989). Currently, there seems to be no direct economic advantage in using mycorrhizal inoculants in high-value crops. However, where the economic value of the crop is low and where considerable amounts of phosphate would otherwise need to be added to the soil, as in the reclamation of acid phosphate-deficient soils of the tropics, then the use of mycorrhizal inoculants may become a practical and economic reality. Even when a good case can be made for the use of mycorrhiza, one needs to determine whether artificial inoculation or manipulation of the population of native mycorrhizal fungi by cropping practices is the best means to the end (Stribley, 1989).

Biological control agents

Microbial inocula are already used worldwide in the control of some diseases and pests in intensive agriculture and there is some scope for their use in less intensive, low-input agricultural systems in developing countries. *Bacillus thuringiensis* is already applied to an extent in developing countries, for example, in the control of pesticide-resistant blackfly vectors of river blindness in West Africa. Research is being conducted in more advanced developing countries, like Thailand and Indonesia.

The production of microbial inoculants is not very difficult; significant quantities can be produced in unsophisticated fermentors of modest volume (Eaglesham, 1989). What is more difficult is the selection of effective strains which show consistent benefits: strains should maintain viability and sustain biological activity. Quality control of the inoculants is very important and requires the development of rapid assays for biological activity (growth promotion or biological control) for use during product development and production. Furthermore, extensive regional trials would need to be conducted with the product to determine the environmental limits on biological activity and monitor the survival and dispersal of the inocula (Davison, 1988). Attention should also be paid to delivery systems to allow application by small-scale farmers.

Plant cell and tissue culture

Plant tissue culture

The basis of plant tissue culture is the ability of many plant species to regenerate a whole plant from tissue or a single cell. Plant tissue culture is a simple and straightforward technique which many developing countries have already mastered. Its application only requires a sterile workplace (a vertical laminar air flow hood), nursery and greenhouse, and (not necessarily highly) trained manpower. Tissue culture is, however, labour intensive, time consuming and consequently costly. In industrialized countries, most plants produced by tissue culture must be sold for more than US$0.25 per plant. Despite lower labour costs in developing countries, tissue culture is likely to be a more expensive option than seed in the production of large volumes of plant propagation material (UNDP, 1989). On the other hand, for the production of early generation, disease-free plants, and some high-value plants in horticulture, tissue culture propagation is being used with much success in several countries. Plants of importance to developing countries that have been grown in tissue culture include oilpalm, plantain, pine, fir, banana, date, eggplant, ginseng, jojoba, pineapple, orchid, rubber tree, cassava, yam, sweet potato and tomato (Sasson, 1989).

Plant cell culture

Plant cell culture involves the production of secondary metabolites by differentiated or undifferentiated plant cells which are cultured in liquid medium. It is a further development of plant tissue culture and is related to fermentation. Plant cells in culture produce a range of compounds which may be of economic interest, like shikonine, quinine, morphine, nicotine, alkaloids, codeine, cocoa, ginseng and pyrethrum. Plant cell culture is a high-cost and labour-intensive technology. It is still a developing technology. Currently, the end-product's value should be greater than US$1000 per kilogram before it can be produced profitably in plant cell culture.

Plant cell culture has rendered few applications in developing countries, so far, and offers little scope in the foreseeable future. Plant cell culture represents more a threat to agricultural production in developing countries (loss of export markets) than an opportunity.

Enzyme technology

Enzymes are proteins that catalyse chemical reactions while remaining unchanged upon the reaction's completion. Part of enzyme technology (that involving the production, purification and immobilization of enzymes) is complicated and most developing countries cannot yet produce high quality enzymes. Their utilization is, however, easy and widespread. Only the more advanced developing countries like Thailand, India, Brazil, Cuba and Mexico are active across the whole spectrum of enzyme technology. Enzyme technology provides both opportunities (import substitution) and threats (loss of export markets) to the economies of developing countries.

Embryo transfer

The costs of embryo transfer (recovery, storage and implantation of animal embryos) are quite high; the hormones involved and often the embryos themselves have to be imported. The application of embryo technology in the field demands access to proper methods for cryopreservation so that embryos collected in an experiment station can be taken to distant places. Costs per embryo transferred have been estimated at US$100-500. In many countries in Asia and Latin America, there have been experiments with embryo transfer in cows and buffaloes; the work in cows has been quite successful. In general, financial constraints will hamper application of embryo transfer in small-scale agriculture.

Protoplast fusion

A protoplast is a plant cell of which the cell wall has been stripped away by enzyme treatment. Protoplasts (often of different plant species) can be made to fuse with one another, resulting in an unpredictable new genotypic composition. The techniques for protoplast fusion can be relatively easily applied in developing countries. There are already quite a number of developing countries capable of executing protoplast fusion. The practical possibilities for solving agricultural problems are, however, limited. In plant breeding, recombinant DNA technology is generally preferred over protoplast fusion (Nijkamp, pers. comm.).

Hybridoma or monoclonal antibody technology

In hybridoma technology, an antibody-producing mammalian B-lymphocyte is fused with a type of cancer cell. The resultant hybridoma cell will produce large amounts of a single antibody, a monoclonal antibody. Developing countries can relatively easily assimilate the relevant techniques for producing monoclonal antibodies. The laboratory facilities required are modest. The cost of the basic equipment is equally modest and will not exceed US$50,000 (Campbell, 1984). However, monoclonal antibody production is laborious and costly compared with the conventional production of polyclonal antiserum. It may take from a few months to a few years to obtain a monoclonal antibody, depending on the expertise of the scientist, the quality of the facilities and the characteristics of the antigen to be isolated (Campbell, 1984). The recurrent costs are relatively high (van Minnen, pers. comm.). For this reason, it is common practice in the developed nations to exchange hybridoma cells between laboratories if possible. While, in theory, this could be extended to developing countries, communication, distribution and financial problems may hamper it.

The equipment for large-scale production is quite specialized. Production in developing countries could start on a smaller scale with simple equipment, especially since cheap labour can be used. A quality laboratory for the testing of raw materials and end-products is extremely important. The same applies to quality control during production. Production of diagnostic kits can take place in developing countries. With cheap labour, unit-price can be as low as US$12-15 per 100-test kit (Vroemen *et al.*, 1989). Very few laboratories in developing countries can, yet, develop and/or produce monoclonal antibodies. In Thailand and India, they are produced on a small scale for research purposes.

Although local production is not an option for most developing countries yet, monoclonal antibody-based tests are of relevance to developing countries in plant and animal health (disease diagnosis) and animal reproduction (detection of hormone levels in female animals). The technology is well developed and immediate application is feasible. Moreover, monoclonal antibody tests are relatively robust, and much simpler to use than DNA probes (see below). The range of this kind of diagnostic test will be widened in the near future to offer a large variety of accurate and easy-to-use kits (Sasson, 1989). One important problem that has to be solved for the application of diagnostic kits in developing countries is their stability in 40°C conditions (Vroemen *et al.*, 1989). Polyclonal antibodies can also be extremely useful, and their broader spectrum can sometimes be preferred to the highly specific monoclonal antibodies (Campbell, 1984; Persley, 1990). The advantages and disadvantages must be assessed on a case-by-case basis.

Recombinant DNA technology

DNA probes

DNA probes (gene probes) are single-stranded DNA fragments (often tagged with a radioactive, fluorescent or enzyme label) which attach themselves to a piece of DNA or RNA which is complementary to the probe. DNA probes, therefore, can be used to look for the presence of specific DNA sequences in organisms or in isolated DNA preparations, which is relevant for the identification, isolation and mapping of genes.

Currently, DNA probe-based tests are quite expensive; each test costs about US$7. In the near future, however, it should be possible to reduce the costs which will enhance application of these tests in developing countries (Van Brunt and Klausner, 1987). For widespread application of DNA probes in developing countries the tests will have to be stable, sensitive and use colorimetric, rather than radioisotopic, labels. Only a few DNA probe-based tests fulfil these requirements as yet. However, as the technique evolves, the number of such DNA probes will increase. Several more advanced developing countries (e.g. Thailand and India) are conducting research on DNA probes.

Transformation of micro-organisms

Recombinant DNA technology has conferred the possibility to change the characteristics of micro-organisms specifically. The application of rDNA technology to micro-organisms involves the identification, isolation, characterization, transfer, cloning and expression of genes. The transformation of micro-organisms is a highly sophisticated technique which requires extensive financial and technical input. It requires well-staffed, fully-equipped and high-safety laboratories. Only very few developing countries can transform micro-organisms at the laboratory level.

At present, only the development of recombinant plant-beneficial bacteria seems of sufficient interest for small-scale agriculture in developing countries. The usefulness of somatotropins to accelerate livestock growth and lactation in developing countries is doubtful. In most cases, financial and technical constraints will determine that other techniques should be adopted before rDNA technology (Sasson, 1989).

rDNA vaccines

Vaccines must achieve active and long-lasting immunity (humoral and/or cell-mediated) to prevent the development of disease upon exposure to the corresponding pathogen. However, vaccines themselves must not be pathogenic. The emergence of rDNA technology has led to the development of a new range of vaccines. Recombinant DNA vaccines are more stable and safer, and can be developed against diseases which cannot be prevented through conventional vaccines. Their development is, however, more costly. This cost is estimated to be US$5-10 million; more than 10 times that of conventional vaccines. The equipment and materials needed are quite expensive and the personnel must be highly qualified. It takes at least five years to construct a candidate rDNA vaccine and another five years to develop that as a marketable product (de Graaf, pers.comm.). Only very few developing countries could currently develop and produce rDNA vaccines. Nevertheless, rDNA vaccines which are safe, effective and heat-stable are of great relevance to developing countries; controlling animal diseases is not only important in improving the health of the animals per se, but also in gaining access to the export markets for live animals and meat.

Transformation of plants and animals

The techniques of rDNA technology to transform plants or animals are basically the same as, though more difficult than, the ones used to transform micro-organisms. rDNA technology in plant and animal breeding makes it possible to cross-breed species that are too far apart for normal sexual reproduction and to incorporate specific characteristics. Furthermore, biotechnological plant breeding can reduce the time-span needed to develop a new plant variety. It takes about 10-20 years to develop a new plant variety through conventional plant breeding while rDNA technology achieves the same result in 5-10 years (Nijkamp, pers. comm.). However, biotechnological plant breeding is much more expensive than conventional plant breeding. The cost of reagents and enzymes is high and they are often difficult to store (UNDP, 1989). Transformation of plants and animals can only be successfully performed in well-staffed, fully-equipped laboratories and most of those are in industrialized countries; only very few developing countries are, yet, capable of this kind of research (e.g. Thailand and India).

It would be too optimistic, therefore, to expect that genetic engineering in plants and animals will improve agricultural production in developing countries in the short term. In the long term, however, the potential benefits, in combating plant viral diseases for instance, are extremely high. In considering rDNA approaches to plant or animal breeding, what has to be done is to compare the feasibility, costs and time involved with conventional approaches.

Conclusions

Most of the developing countries cannot themselves apply rDNA technology or hybridoma technology in research programmes. The basic disciplines involved have not yet been mastered and this makes any attempt at integrated application very difficult. High technology facilities need specialized maintenance and constant attention. Both the researchers themselves and the maintenance personnel need to be highly trained. At present, equipment and spare parts must in the main be purchased abroad. rDNA technology is characterized by the use of large amounts and many different types of chemicals. Their supply and storage cause problems; delivery times are long and shortages of hard currency might arise at any time. Some of the chemicals and enzymes must be transported and stored in a frozen state (restriction enzymes), others have to be transported relatively quickly given their short half-life (radio-isotopes). Researchers in developing countries depend on industrialized countries not only for the necessary reagents but also for the smooth turning of the wheels of bureaucracy, something that can only rarely be guaranteed (Vroemen *et al.*, 1989).

Selected university laboratories in more advanced developing countries and international research centres will be capable of carrying out the work in collaboration with scientists in industrialized countries. But even in such special situations, there will most likely need to be extensive financial and technical input from industrialized countries if rDNA technology is to have an impact on agricultural production in developing countries.

Developing countries that take the decision to use genetic engineering to help solve agricultural problems will have to decide whether to develop the necessary systems internally, use available systems from industrialized countries or seek partners in industrialized and/or more advanced developing countries to assist in development of specific biotechnologies. At present, it appears more efficient for most developing countries to use existing systems, adapting them to local needs if necessary (Persley, 1990).

The environmental release of genetically modified organisms (GMOs) is a very important issue. Risk assessment of GMOs is still in its infancy. While research to improve risk assessment is being conducted, many industrialized countries have instituted strict regulations on the handling of GMOs in the laboratory and the environment. There are moves towards the liberalization of these rules in some countries with respect to certain classes of organisms and certain types of genetic modification which are now the subject of notification, rather than assessment, procedures. Nevertheless, risk assessment is still largely conducted on a case-by-case, step-by-step basis. Developing countries have no legislation on GMOs at the moment, although there are safety committees in some developing countries like India. Even if these countries had stringent regulations it is doubtful whether they could police them without extensive staff training and a much greater experience of risk assessment of GMOs.

For fermentation technology, microbial inoculation of plants and plant tissue culture, the requirements are less strict. The techniques used are less complex and more easily transferable. Developing countries are already involved in several types of fermentation especially solid state fermentation (SSF), alcoholic fermentation and biogas production. Also more sophisticated fermentation technology has been successfully introduced in several developing countries. Microbial inoculants, especially rhizobia inoculants, are already produced in many developing countries. Biological control agents have not yet been produced, but this would be plausible. Many developing countries have also mastered plant tissue culture, though only on the laboratory level, rather than on a commercial scale. The more advanced developing countries in Asia and Latin-America also have the capability to perform embryo transfer (ET) and protoplast fusion. Enzyme technology (production, purification and immobilization of enzymes) is, however, more complicated. Most developing countries cannot yet produce high quality enzymes, although enzyme use is widespread.

Biotechnology offers several interesting opportunities of solving certain problems in crop production at the small-scale farm level in developing countries. In the short term, the major applications of biotechnology are plant tissue culture, improved biological nitrogen fixation, biological control agents, and improved diagnostics for plant and animal diseases. The major benefits to agriculture from genetic engineering of plants (e.g. improved resistance to diseases and insect pests, and improved nutritional quality) are likely to emerge after the year 2010, although some applications will be realized earlier. Such applications seem to be well suited to the needs of sustainable, low-external-input agriculture.

In livestock production, biotechnology could contribute to the solution of several problems on small-scale farms: (i) the improvement of animal nutrition through the upgrading of agricultural and agro-industrial wastes, and through improved fodder crop production; (ii) the improvement of animal health with better diagnostics and vaccines; and (iii) the enhancement of animal reproduction through hormone monitoring in female animals. In developing countries, cattle (in Asia, also buffaloes) will be the primary target for the biotechnology. They are the most important species, economically and socially, and their high individual value justifies the costs of new biotechnological inputs (Persley, 1990). It must be noted, however, that in the case of livestock production the results of biotechnology, unless subsidized, are often likely to be more useful to producers already using high technology than to those operating less intensively.

In local processing of agricultural raw materials, biotechnology could increase farmers' ability to upgrade the quality of and preserve their products. In the short term, for instance, 'solid state fermentations' for the production of food and feed, or biogas production developed for the generation of energy and treatment of agricultural wastes could be improved.

1.3   Current developments and impacts

Despite its great potential value for small-scale farmers in developing countries, few relevant biotechnological research and development projects have been executed. To find the reason for that, it is necessary to examine what might be called the 'resource base' of biotechnological innovation - the research and development programmes of the rich, developed nations.

The problems of small-scale farmers rarely feature on mainstream agendas either of public or private R&D organizations. If this situation is to be altered, one needs to examine the origins of those agendas. Who compiles them and who influences their content? The most active groups in biotechnology research are government institutions (universities and public research institutes) and private companies (research firms and multinationals in the chemical, agrochemical, pharmaceutical and food sectors). Much of the fundamental molecular biological and genetic research underlying modern biotechnology has been developed in governmental institutions. The private sector, too, has become interested in biotechnology. Companies, large corporations in particular, have rapidly built ties with universities and public research institutes. Waning government research funding in many developed countries has put increased pressure on universities and public research institutes to engage in contract research. Research contracts add exclusivity of access to public sector research and provide additional input for private sector development of products. Through contract research and the acquisition of biotechnological research firms, several multinationals have attained a dominant position in biotechnology, a position protected by intellectual property rights (patents) (Broerse, 1990; Clark and Juma, 1991; Junne, 1987; Kenney, 1986).

Governments are also heavily involved in biotechnology. In most industrialized countries (e.g. USA, Japan, United Kingdom, France, Germany and the Netherlands) the development of biotechnology has high priority. Governments endeavour to stimulate and influence developments in biotechnology through grants both to industry and public institutions. The priorities addressed by these programmes usually stem from market analyses which identify the problems and interests of industry.

Thus, current biotechnological R&D programmes are, by and large, guided by the economic considerations of industry (and governments). Unsurprisingly, the type of technology that will ultimately emerge and the products that will be developed are similarly oriented. Biotechnology is, therefore, primarily used where it can lead relatively easily to commercially attractive applications for which large and lucrative markets exist, e.g. diagnostics (immunological tests and DNA probes), human pharmaceutical and animal vaccines, plant improvement (the addition of single gene traits such as herbicide- or pest-resistance to hybrid seeds), and food processing (effluent treatment, 'natural' additive production and enzyme-based processing) (Broerse, 1990).

Although small-scale farmers in developing countries have virtually no access to mainstream biotechnological research, they are nevertheless intimately affected by its application by others. For them, mainstream biotechnology is mainly negative. Two drawbacks, in particular, should be mentioned.

The industrialization of agriculture

Agricultural research generally is focused on increasing the yield of cash crops: the development of high-yielding varieties is combined with the supply of labour-saving and capital- and energy-intensive input packages (e.g. fertilizers, chemicals and mechanization). This research orientation often brings little reward for small-scale farmers in developing countries; the resulting innovations are not suited to the marginal conditions with which they contend and, in any case, most cannot make the necessary investments (Broerse, 1990; Junne, 1987). In contrast, resource-rich producers in developing countries can usually adopt the new innovations rapidly. Consequently, their output increases while that of small-scale farmers remains virtually the same. Increased supply depresses commodity prices, forcing many small-scale farmers into an even more marginal position: any excess production that they might have had is then worth less on the open market. Biotechnological research in agriculture is, in the main, no different from previous research. Genetic engineering is simply another tool in the development of higher-yielding crops and more cost-efficient production. Its expected consequences for the small-scale farmers - the erosion of cash markets - are, therefore, likely to be similar.

Interchangeability of raw materials

Biotechnology enhances the interchangeability of raw materials (Broerse, 1990; Clark and Juma, 1991; Junne, 1987; Roobeek, 1987; Ruivenkamp, 1989). It introduces functionally-identical, novel substitutes for traditional raw materials (e.g. sweeteners for sugar). It also permits the production of existing raw materials in industrial rather than agricultural settings: for instance, pyrethroids, vanillin or the purple dye shikonin can be produced through plant cell culture. Biotechnology could also alter the markets for tropical commodities even in the absence of such a radical innovation. For instance, crops like coconut and oilpalm are, by their nature, interchangeable in the production of vegetable oils: any improvement biotechnology made to the yield or quality of oil from one of them would change the supply considerably. Thus the processing industry (mainly located in industrialized countries) could select the most attractive raw material for its products. The fragile economic position and vulnerable farming systems of small-scale farmers mean that these farmers cannot react to such changing market demands. The result - many small-scale farmers will be further marginalized. The best-known example - the replacement of sugar (saccharose) by another sweetener (high fructose corn syrup) - demonstrates the speed and severity of the impact of commodity substitution on Third World agriculture (Box 1).

1.4 Understanding the gap

Clearly there is a large and widening gap between the small-scale farmers' need for solutions to their problems (what economists might call 'the demand') and the supply of innovations by scientists and others. Before we can hope to close the gap, its nature and origin need to be defined. On the *demand side*, we find small-scale farmers who have much knowledge of their environment and experience of how to use that environment. They are not, as might be assumed, inately conservative. On the contrary, they are usually active and creative in experimenting with local innovations (ILEIA, 1990). However, largely bereft of purchasing power, external innovations of potential benefit do not reach them through the commercial process. Dispersed, isolated and poor, their influence on the political agenda, even within their own countries, is minimal (World Bank, 1990). Their isolation to a large extent denies them exposure either to information from formal agricultural science (Biggs, 1989) or to practical innovation purveyed through the less formal farmers' days. Their opportunities for articulating their problems and needs are similarly limited (Chambers and Ghildyal, 1985).

Box 1: The impact of the increasing interchangeability of raw materials on Third World agriculture

In 1965, Japanese biotechnologists developed a bacterium which produces the enzyme glucose isomerase in high yields. This opened the way to the commercial production of high fructose corn syrup (HFCS) from crops like maize, wheat, or potato. As a result of the introduction of HFCS in soft drinks, sugar consumption in the USA decreased from 10.8 million tonnes in 1979 to 8.6 million tonnes in 1984. In 1984, the Coca Cola Company and Pepsico disclosed their intention to substitute all sugar by HFCS in soft drinks within 2 years. The market price of raw sugar collapsed from around US$0.62 in 1980 to only US$0.12 in 1985. Both companies did not offer any compensation to their traditional suppliers to enable them to convert to the cultivation of alternative crops. The impact of this biotechnological innovation was worst on Negroes in The Philippines, an island which depended for 90% of its export on sugarcane. The sugar export earnings in the Philippines dropped from about US$650 million in 1980 to about US$250 million in 1985 (Ahmed, 1988; Ruivenkamp, 1989).

If biotechnology proceeds in its current direction, and it shows every sign of doing so, it will surely by-pass small-scale farmers and deprive them of any benefits it might have brought. Furthermore, inasmuch as it promotes agricultural intensification, and raw material exchange, it will erode their access to capital markets and thereby stunt their efforts to progress from subsistence to stability.

On the *supply side* of innovation, the scientists and technologists are as isolated as the small-scale farmers. Their innovative energies are channelled into the agricultural system to which they are both geographically close and infrastructurally bound - an agriculture which is both capital- and energy-intensive (Bunders, 1988; Wolf, 1986). The gap between the supply of technological innovations and the small-scale farmers' demands cannot be satisfied by the much-trumpeted 'trickle-down effect' from the resource-rich farmers to the resource-poor, small-scale farmers. The phenomenon rarely occurs, largely because the small-scale farmers cannot afford the costly materials and services associated with these innovations (Bunders, 1988). Even when 'trickle-down' does occur, it can have unfortunate and unforeseen side-effects.

In sum, we have an unarticulated and weak demand for technical solutions to small-scale farmers' problems and needs, and on the supply side we have a research orientation which is unattuned to the diverse, complex and risk-prone conditions within which small-scale farmers strive to produce. Without resolving this mutual isolation, we have no real hope of directing the necessary research. Without an appropriate and thorough design, projects and programmes for small-scale farmers will exact high social and environmental costs. Unless tomorrow's innovations are more consistent with small-scale farmers' needs we will do little more than repeat the injustice and damage of earlier technological revolutions.

1.5   Closing the gap

There are two major problems in closing the gap. Firstly, the absence of a mechanism for an appropriate and thorough design of biotechnology projects and programmes for small-scale farmers. And secondly, the mutual isolation of farmers and technology innovators and its corollary - the insufficient reflection of small-scale farmers' needs and interests in biotechnology research.

Project and programme design

On the supply side of innovation for small-scale agriculture, some progress is being made. Several research institutions in both industrialized and developing countries are to an extent generating ideas and initiating biotechnology projects with small-scale agriculture in mind. Some governments of developing countries and donor organizations, including the Netherlands Directorate General for International Cooperation (DGIS), have established or are formulating biotechnology programmes which offer the possibility of funding biotechnology projects directed to small-scale farmers.

Nevertheless, there is still ample reason for caution. The conventional orientation of agricultural research means that biotechnology projects, even those aimed at a target group of small-scale farmers, are likely to fail. Or, worse, they may have a negative impact on farmers, their environment and the national economy. The vast potential of biotechnology to address many aspects of human life is, at the same time, both a blessing and a curse. Biotechnology is a fast-developing field and its impacts can be far-reaching. Although this can be an advantage (small-scale agriculture requires large and immediate appropriate solutions), the accelerating pace of biotechnology leaves little time to anticipate possible negative consequences or to remedy them once they are apparent.

Institutions which generate and/or fund biotechnology R&D for small-scale farmers face the same dilemma as researchers and research-funding bodies which operate in other areas. They have to choose which of very many proposals to support and pursue and which to ignore. For instance, industry and those investing in industry apply criteria rigidly to select innovative projects, mindful of strengths and weaknesses, aware of past successes and failures, and sensitive to the cost-effective use of resources. In contrast, the use of criteria for assessment of project proposals by development organizations is a rarity. Even when criteria are used, they are usually insufficient to assess the appropriateness of biotechnological innovations for small-scale agriculture in developing countries; they are too vague and do not take into account the conditions under which small-scale farmers produce.

Novel criteria are needed to assess both the appropriateness and feasibility of biotechnological innovations for small-scale agriculture. They are crucial to reduce the chances that money and time are wasted and/or that projects will have significant negative effects. In Part One, we present and discuss new guidelines for the more stringent assessment of research and development proposals.

Directing biotechnological research

Criteria for the judgement of research proposals alone are clearly insufficient. At most, criteria influence organizations which already design and/or fund biotechnology projects and programmes for small-scale farmers in developing countries. To ensure a broad and structural focus on small-scale farmers' needs, it is not sufficient to address only these organizations. What is needed is that governments, research institutes, donor organizations and other providers of support incorporate the research priorities of small-scale farmers into their general (biotechnology) policies.

There are many organizations and social groups involved in bringing innovations to small-scale farmers: scientists, expert consultants, donor organizations, policy-makers, farmers and the organizations which represent and/or work with farmers. Often, all have different perceptions of the problems of small-scale farmers and what constitutes an appropriate solution. Moreover, while each group possesses specific expertise, it lacks other necessary knowledge. The lack of exchange of information impedes any consensus. Those closest to the problems - the small-scale farmers - cannot influence the direction of research. Those who make the decisions - the scientists, industrialists and civil servants - are, in general, the least familiar with the specific problems and needs of small-scale farmers.

We will argue that an approach which enhances information exchange is needed. It must identify the specific needs and problems of small-scale agriculture and must prioritize biotechnological research accordingly. Such an approach should enhance the incorporation of these priorities in biotechnology research agendas. In Part Two, we describe the 'interactive bottom-up approach', a model we have developed and applied to assess the use of biotechnology for small-scale farmers in developing countries. The interactive bottom-up approach avoids technology-push by drawing on the knowledge and opinions not only of scientists, policy-makers and expert consultants, but also that of end-users and the organizations which represent and/or work with them. One of its central themes is the use of two different but closely cooperating teams: a formal interdisciplinary team to bridge the gap between providers of innovations and the potential users, and an informal team on the spot, consisting of people sharing the same commitment, to specify and broadly justify the ideas.

PART ONE

TOWARDS CRITERIA FOR ASSESSMENT OF PROJECT PROPOSALS

Jacqueline E.W. Broerse
Steen Joffe*
Joske F.G. Bunders

*Steen Joffe, Natural Resources Institute (Overseas Development Administration), Central Avenue, Chatham Maritime, Kent ME4 4TB, United Kingdom.

Chapter 2

A case study: yam tissue culture in the Caribbean

2.1 Introduction

Bridging the divide between small-scale farmers in developing countries and the research pool from which they might derive benefit requires ultimately that the limited funds available for development are well spent. Faced with widespread criticism of previous projects and programmes, donors' policies are being questioned, not least by the donors themselves. We believe that donors should take more account of and be more accountable to those who are, in effect, the end-users of their services. To this end, we propose a set of guidelines to aid donors in deciding which projects to support and how to shape them. Basically we hope that these guidelines will support effective and efficient intraction between proposing and funding agencies. Before introducing these guidelines (Chapter 3), however, it will be instructive to examine the kind of context within which they must operate.

For this purpose, we have documented the genesis, evolution and eventual demise of a yam tissue culture project in the Caribbean. The case study illustrates how donor-funded biotechnology projects can have impacts on small-scale agriculture. It is also an example of the sort of development that has led us to propose novel criteria for the assessment of project proposals.

The information we present is based on available published material from the project and on conversations with some of the key participants. All Caribbean case study materials were provided by Dr Sinclair Mantell, Unit for Advanced Propagation Systems (UAPS), Wye College, UK. The information from 1984 onwards was taken from CARDI annual reports and personal communication in September 1990 with Dr Roger Bancroft, plant pathologist in charge of CARDI Tissue Culture Unit since 1986.

The project in brief

During the 1960s the increasing incidence of internal brown spot and several other diseases was having serious effects on yields and quality of yam tubers in Barbados and other countries. This created shortfalls in yam production for indigenous food and problems for the growing export trade.

Yam is a very important subsistence crop in the Caribbean. It is grown mainly by small-scale farmers on plots of less than 0.5 ha and contributes as much as 50% of the dietary calories in the region. The crop is also an important source of export revenue and a feedstock source for local industrial processing.

In 1973, in response to the disease problems, a project using the new techniques of tissue culture started. Tissue culture was to be used to try to eradicate disease and to multiply virus-free seed tubers for distribution to local farmers. The project initially had the backing of the UK Overseas Development Administration. Its early results were very promising and, with the support of other donors, the project's objectives expanded to include the establishment of a self-financing yam tissue culture and propagation laboratory (the 'Tissue Culture Unit') in Barbados under the aegis of the Caribbean Agricultural Research and Development Institute (CARDI). By 1982, disease-free clones of the popular species, *Dioscorea alata*, were available to farmers through an 'Approved Growers' scheme. Average yield gains at farm level were ca. 30-40% and the yam tubers were of consistent high quality; yield gains as high as 95% were reported from initial trials in low technology small-scale farmer production in St Lucia.

After the project ended in 1984, however, things started to go badly wrong. Approved growers took to multiplying the improved varieties conventionally in the fields rather than returning to the Tissue Culture Unit for new disease-free stock. This, together with a number of other factors, may have contributed to a serious outbreak of the yam foliage disease 'anthracnose' during the mid-1980s. The yam variety originally chosen for the tissue culture work seemed especially susceptible to anthracnose. At the same time, the viral disease which the original project had been designed to eliminate reappeared. Because of these difficulties, the Tissue Culture Unit never achieved its objective of self-funding.

In order to be more analytical, and to draw out the reasons why the project successively succeeded and failed (and it did both), one needs to look more closely at the problems it sought to solve, at the technical solutions that were proposed and at the mechanisms through which decisions were reached.

2.2   Background

Importance of yam

Together with cassave and sweet potato, yam belongs to the most important of the world's tropical root and tuber crops. Western Africa accounts for about 90% of world yam production, and the average contribution of yam to the human diet in this region was nearly 300 calories/day between 1975 and 1984 (Gebremeskel and Oyewole, 1987). Throughout the Commonwealth Carribean, yams (*Dioscorea* spp.) are cultivated as a subsistence crop, forming a significant part of the carbohydrate diet there; a 1970 food survey on carbohydrate consumption in Barbados revealed that yams contributed ca. 56% of total calorie intake (CARDI, 1979). They are usually grown on smallholdings of 0.5 ha or less mainly for home consumption, although in some countries, particularly Barbados and Jamaica, most yams are grown on a larger scale of up to 10 ha for commercial purposes (CARDI, 1981). Fresh yam tubers are chipped, dried and packaged to create 'instant yam' flakes which provide a year-round source of carbohydrate for both local and export markets (Mantell and Haque, 1979).

With the expansion of immigrant populations of West Indian origin in Europe and the USA during the 1960s, a significant yam export trade developed: by 1968, the total weight of yam tubers exported from Barbados alone, for instance, was approximately 1000 tonnes (CARDI, 1979). The growing export trade and the food import substitution policy in Commonwealth Carribean countries stimulated the production of local yam cultivars.

Yam production constraints

Diseases

Yams were badly hit by several diseases during the 1960s (Box 2) (Mantell and Haque, 1979).

Box 2: Diseases of yam

Causative agent	Symptoms	Results	Control
Tuber-borne diseases:			
Viruses: internal brown spot, flexous rod-, bacilliform-, and spherical-shaped-viruses	Leaf symptoms: mottle/vein banding and vein clearing Tuber symptoms: internal brown spot	Substantial loss in tuber yield (29-37%) and quality	Cannot be controlled by applications of chemical sprays
Nematodes: dry rot	Extensive tuber rotting		Dipping of affected tuber setts in nematicide solutions before planting Propagation from yam plant vines (uninfested)
Fungi and bacteria:	Extensive rotting of central tuber tissues during the storage phase		Careful handling Curing of cut surfaces followed by applications of chemicals
Yam foliage diseases:			
Air- and rain-borne fungi: leaf spot, and anthracnose	Extensive defoliation of yam vines	Substantial yield losses	Applications of certain conventional organic fungicides

In the mid-1960s shipments of *Dioscorea alata* yams from Barbados and other exporting countries were suffering severely from internal brown spot (IBS). IBS is a disease characterized by the presence of small, brown necrotic spots in the flesh of the tubers detectable only when the yams are cut open. This caused serious problems for the burgeoning export trade. In 1965, 50% of all yam shipments from Barbados were lost. Virus diseases became the limiting factor on commercial production of *D. trifida* yams in Guyana. Dry rot infestations caused by nematodes severely constrained yam production in Jamaica.

Epidemiological research in the 1970s revealed that virus diseases affected all yams in the Caribbean. Conventional disease control methods had proved inadequate while traditional propagation techniques simply exacerbated the problems. Of particular concern was the prevalence of IBS; it was found to be widespread in *D. alata* (the most popular species of yam) in Jamaica, Dominica, St Lucia, Barbados, St Vincent, Grenada and Trinidad. The infection level varied from island to island with the highest (71%) being found in Barbados and the lowest (32%) in Dominica.

Agronomic problems

There are other constraints too in Caribbean yam production. The most notable is the wide phenotypic variation and irregular shape of tubers of cultivated stocks (CARDI, 1987). For commercial use, export trade and industrial processing, there is a need to improve the uniformity and tuber shape of *D. alata* yams.

2.3 Yam virus project (1973-84)

Research phase (1973-79)

In response to the virus infection problems, a yam virus research project started in 1973 at the Regional Research Centre (RRC) in Trinidad. It received three years of funding from the UK Overseas Development Administration (under ODA Scheme R2672). Its objectives were (CARDI, 1979):

1 . To determine the mode of transmission of yam viruses.
2 . To identify the causal agents of internal brown spot (IBS) disease.
3 . To select disease-free clones from farmers' stocks.

4. To develop methods for the eradication of virus diseases in commercial stocks.

5. To measure the effects of virus diseases on tuber yield.

Work was carried out in Trinidad (greenhouse and laboratory studies) and Barbados (field studies). In 1975, after the project had been assessed by Dr B.D. Harrison, a virologist at the Scottish Horticultural Research Institute, a proposal for an extension of the project was made to the ODA by the successor organization to the RRC, the Caribbean Agricultural Research and Development Institute (CARDI). With funding secured, studies continued for a further three years (under ODA Schemes R3232 and 3218) finishing only in 1979.

Application of biotechnology

Following a consideration of the technical options, the researchers believed that it should be possible to address simultaneously both IBS and the agronomic problems associated with yam through tissue culture propagation. Research started in 1973. Dr Sinclair Mantell, an ODA-funded UK tissue culture expert, and his colleagues established the first successful system for eradicating yam pathogens by what was then the relatively new technique of tissue culture. The *D. alata* cultivar variety (c.v.) White Lisbon was selected for this work because it was one of the most important in terms of total area under cultivation and marketability, both locally and as an export product. It was also the variety of choice for the production of 'instant yam flakes'. Two tissue culture techniques were applied:

1. Meristem tip culture: Apical meristem tip culture involves removing the small growing tip (apical meristem) of the yam vine and culturing it under aseptic conditions in a glass tube on an artificial growth medium. The growth medium contains essential minerals in the proportions required for healthy growth; plant hormones are added to induce shoot and root formation. In this way, the meristem tip can be regenerated into a new yam plant. Since viruses are usually absent or, at least, present only at greatly reduced concentrations in the meristem tip, the regenerated yam plant is free, or virtually free, of pathogens (Mantell and Haque, 1979).

2. Micropropagation: In micropropagation, the small, disease-free plantlets produced by apical meristem tip culture are propagated further. Small pieces of stems of the plantlets are subdivided into microcuttings which are grown in much the same way as the apical meristems to produce more shoot material. Through micropropagation, 65,000 plants can be obtained from a single parent in a 6 month period. As the work takes place in laboratory conditions, multiplication can be undertaken at any time, even during the yam dormant season (Mantell and Haque, 1979).

Before the plants are ready for the field, they must undergo two further production stages. Firstly, they have to be 'hardened off'. Once the young plants have developed root and shoot tissues, they must be established in a sterilized potting mixture (composed of soil, sand and an organic supplement such as peat moss or finely ground bagasse) in an insect proof greenhouse. After they have matured sufficiently in the greenhouse, they are transferred to soil, at an isolated field site. A number of precautions are taken in the field to prevent re-infection. For instance, the site is located upwind of areas under yam cultivation and the plants are grown in 'gauzehouses' to protect them from flying insects. The tubers produced at this site are the 'elite stock'.

Outputs

The 'elite' planting material produced through tissue culture in the research phase of the project was available in three grades, 'A', 'B', and 'C', distinguished by the degree to which they were free of pathogens. C grade material, the least virus-free, was available in relatively large quantities. However, A grade stock (the most pathogen-free) was still in tissue culture at that time. Consequently, it was decided to begin multiplication of C and B grade stocks, with the intention of progressing to A grade material as soon as that was possible.

In 1979, seventy-three accessions (plants derived from a single source) of B and C grade material were planted on isolated (CARDI) plots in Barbados. Their growth was described as 'extremely rapid' and the resultant total foliage as 'markedly outstanding'. Only two plants of the 73 showed severe disease symptoms; all others either exhibited only mild symptoms or were apparently free from disease. Their average yield overall was ca. 24 tonnes/ha, with the highest yielding type reaching ca. 31 tonnes/ha. At the time, average yields in the Caribbean of 'conventional' yam stocks were ca. 15 tonnes/ha.

Implementation phase (1980-84)

After the experimental phase, it was clear to the researchers and Dr Harrison who assessed the project again in 1978, that the production of disease-free yam tubers by tissue culture was technically viable. It was clear to them, too, that the potential benefits to farmers were substantial; yield, quality and consistency of production were all increased following tissue culture. As a result, CARDI formulated another proposal which encompassed not only an extension of the multiplication and field trials, but also the establishment of a purpose-built micropropagation unit in Barbados. In brief, the objectives were:

1. To provide substantial quantities of tuber planting material of pathogen-tested *D. alata* and *D. trifida* clones to supply CARDI field units, Barbados yam growers and ministries of agriculture for multiplication and distribution to large and small yam farmers in the Commonwealth Caribbean.
2. To establish a nuclear, tissue culture stock of selected virus-tested *D. alata* and *D. trifida* clones which could be used for future propagation.
3. To monitor the rate of re-infection of virus-tested yam stocks by virus diseases when grown under commercial yam production systems.
4. To carry out a cost/benefit analysis of use of pathogen-tested yam seed under different yam production systems.

Donors again responded positively to the proposal. The European Development Fund through the Caribbean Development Bank funded the project between September 1980 and March 1984.

To facilitate the first of those objectives - the provision of substantial quantities of virus-free tuber to local growers - an 'Approved Growers' scheme was devised. Selected 'Approved Growers' in Barbados could purchase 'elite' planting material from the tissue culture unit and re-sell their output of yams to commercial and subsistence growers as 'Certified Disease-Free' seed tubers. A precondition for 'Approved Growers' was that they adhere to a stringent set of conditions and follow precise instructions for growing the materials. A wide range of constraints were specified including the site location, planting and cultivation practices, applications of fertilizers and other inputs, harvesting and storage techniques. Furthermore, all the yams produced were to be sold for planting purposes only. Thus, disease-free planting material was to be produced by farmers for farmers. It was envisaged that the Tissue Culture

Unit could recoup its operational costs from sales of 'Elite Stock' while 'Approved Growers' would profit from sales of seed tubers at premium prices justified by the yield increases (Mantell and Haque, 1979).

The scheme was fleshed out in 1979 with some projections of the potential profitability of a yam tissue culture and propagation unit (Table 1). The figures in Table 1 are illustrative rather than empirical but are based on reasonable (1979) estimates of operational costs (salaries, services, etc.), the productivity of the propagation process, and the price which 'elite' planting materials could command (CARDI, 1978). It was assumed that the capital costs and first three years' recurrent costs (Table 2), which had been worked out on the basis of the earlier research programme would be met by external agencies (CARDI, 1979, 1981).

Table 1: Estimated operational costs in Barbados dollars of a yam tissue culture unit

Costs (per annum)	BDS $
Salaries	28,000
Research Materials	10,000
Travel	6,000
Services	6,000

	50,000
Income (per annum)	
Sales of Elite Stock (18.55 tonnes at $2.7/kg)	50,085
Profit Margin	**85**

Table 2: Resources and costs in establishing and running a tissue culture unit. (All costs are the 1979 original estimates, except the laboratory capital costs which is a contractor's quotation from 1981.)

A. Capital Items

  (a) Tissue Culture Laboratory and Yam
      Propagation Centre
      (including : office, tissue culture lab.,
      potting and sterilization rooms,
      greenhouses and screened
      propagation area)   ca. £70,000

  (b) Other Capital Items
      (including : furnishings, mist irrigation
      system, culture room lights,
      air conditioning, autoclave, refrigeration,
      water still,soil sterilizer, etc.)   ca. £10,000

  Total Capital Items: ca. £80,000

B. Recurrent Costs (per annum)

  (a) Salaries (full- and part-time)
      (including : agronomist/horticulturalist,
      propagation officer, technical assistant,
      lab. attendent, consultant supervisors)   ca. £25,000

  (b) Other Recurrent Costs
      (including : rent,[1] culture materials, service
      charges, labour for weeding/harvesting etc.,
      transportation and storage, publishing
      bulletins, travel, electron microscopy)   ca. £10,000

  Total Annual Recurrent Costs   ca. £35,000

TOTAL CAPITAL AND RECURRENT COSTS OVER 3 YEARS
(80,000 + 3 x 35,000)   ca. **£185,000**

[1]The required land of 0.25 ha was made available under an arrangement with the Barbados Sugar Producers Association for a nominal rent of BDS$5.00 per annum.

Several organizations were involved in the implementation of the second phase of the yam tissue culture project (CARDI, 1979, 1981):

1. The Caribbean Agricultural Research and Development Institute (CARDI) provided technical expertise, administrative facilities and some capital and recurrent cost inputs.
2. The European Development Fund provided BDS$700,000 (ca. £175,000) funds for capital and recurrent costs under the EDF Regional Programme for the Caribbean (Project No. 4100 033 97 46, yam tuber production; consultancy contract No. EDF R/5).
3. The Caribbean Development Bank administered the EDF funds.
4. The Barbados Ministry of Agriculture strongly supported the scheme and contributed (without charge) consultancy and technical services, 3 acres (1.2 ha) of land, labour for cultivations and irrigation, and assistance with transport and storage of yam seed. In return, the project provided training in tissue culture techniques to 4 technicians from the Ministry (as well as one from St Lucia and one from Dominica).
5. The Barbados Sugar Producers Association (yam is often intercropped with sugarcane) gave political support as well as land (at a nominal rent) on which to build the tissue culture unit, labour to assist with field propagation, and the use of tillage equipment. It also assured the cooperation of commercial yam growers in the scheme.

The overall coordination of the project (1980-84) was the direct responsibility of Dr Syed Haque of CARDI, who undertook some of the original research on yam tissue culture with Dr Sinclair Mantell. The senior scientist in charge of the yam propagation laboratory was UK agronomist Frances Chandler (CARDI, 1979, 1981, 1982).

In order to foster a collaborative spirit among the many participating organizations, an advisory committee with representatives of each was established. The committee acted as a channel of communication between the agricultural organizations involved in the scheme and the scientists engaged in propagation work. It was proposed that chairmanship of the committee be rotated between the members.

Outputs

Multiplication of elite stock by approved growers The fruits of the venture were not long in coming. Yams selected from isolated plots were provided to 6 Approved Growers (all members of the Barbados Sugar

Producers' Association) for planting in May/June 1981 (CARDI, 1981). At this stage the areas under cultivation were still less than 1.0 ha per grower, totaling 4.4 ha. By May 1983, 850 kg of A grade material was available to Approved Growers. Virus-tested material was distributed to small- and large-scale growers in Barbados and to CARDI units, ministries of agriculture and other agricultural organizations in the region. The countries to which the material was shipped were Antigua, St Lucia, Grenada, Montserrat, Jamaica, Trinidad and Tobago, Cayman Islands, St Croix, St Vincent, Dominica and St Kitts. The hectareage of virus-tested material grown in Barbados increased from 5 ha in 1981-82, to 20 ha in 1982-83, and 70 ha in 1983-84 (CARDI, 1984). As a result of the yam field day held in 1982 (see Mechanization of yam planting and harvesting), a number of growers either completely mechanized their cropping system or at least harvested it mechanically from 1983.

Restoration of export markets   The export market for Barbados yams showed signs of recovery almost as soon as the first disease-free tubers were in the ground. A market was secured in Britain, for example, for ca. 230 tonnes of yam from the 1983 harvest. The general objective was to export between 30 and 35% of their total output. To expand exports, markets in the USA, Virgin Islands, Martinique and Guadeloupe, and other regional states were assessed.

Mechanization of yam planting and harvesting   With the availability of virus-tested material to improve yield and quality, the prospects for the successful re-establishment of yam as a large-scale export crop looked rosy. To accelerate progress further, by reducing production costs, it was considered that mechanical planting and harvesting practices would have to be developed and adopted. Traditional yam harvesting is a tedious, time-consuming and labour-intensive operation which creates a labour shortfall at the start of the sugarcane harvest. It was, therefore, proposed that the traditional practice of intercropping yam with sugarcane would be replaced with a yam monoculture system to facilitate mechanized production. At the same time, methods for 'suitable' land preparation and accurate planting would have to be developed (CARDI, 1983).

Farmers were trained in these practices through demonstrations on field days. The biggest one of these was held in January 1982 by CARDI in conjunction with the Barbados Sugar Producers Association, on the land of one of the Approved Growers. Its purpose was to educate local yam farmers (ca. 75 attended) in the recognition of disease symptoms and in the new methods which had become available for eradicating them. Another purpose was to promote a more profitable system of yam production - the use of virus-tested material in combination with mechanical planting and harvesting.

At the field day, several farmers expressed concern that their land, which used to produce two crops when intercropped, would now produce only one. The 'solution' demonstrated at the field day was to plant (mechanically) a 'forced-back' sugarcane crop immediately after the yam harvest. This is especially feasible, it was suggested, in high rainfall or irrigated areas. It would be feasible in low rainfall areas too, if the planting operation was carefully planned to take place immediately after yam harvesting, before the soil moisture was lost (CARDI, 1983).

Completion of yam tissue culture and propagation laboratory   The new laboratory and associated facilities were completed in June 1982 and officially opened on 22 October 1982 by the Barbadian Minister of Agriculture. By this time, techniques similar to those successfully employed for the *D. alata* yam, had also been used on *D. trifida*, the major food yam of Guyana. A number of cassava cultures received from CIAT, too, had been multiplied with 'some' success.

Performance of certified stock in small-scale farming systems   At the same time as the original B and C grade elite planting material was being distributed to Approved Growers in Barbados, some of this stock was shipped to the CARDI unit in St Lucia. There, yam production is almost exclusively the province of small-scale farmers with low levels of technology (unlike in Barbados). Input usage is minimal and productivity low, but yam contributes as much as 40% of total farm income. After multiplication by CARDI, certified tubers were distributed to selected farmers to be grown alongside their own varieties according to their normal practices. This, then, was likely to be a much more severe test of the appropriateness of the new crops. An evaluation of the performance in St Lucia of the certified tubers was undertaken by CARDI (Table 3). The figures in Table 3 represent average results from eight different farms; cultivation practices varied from farm to farm, as did the cropping systems and the climatic and soil conditions (CARDI, 1984).

Table 3: Partial budget of on-farm yam production tests (small-scale farming system - St Lucia)

	Farmers' Yam	Virus-Tested Yam
Yield		
Mean Yield (tonnes/acre)	14.18	27.59
Losses (poor handling/storage, etc.)	- 1.41	- 2.77
Net Yield (tonnes/acre)	12.77	24.82
Gross Revenue ($1000/tonne)	12,770	23,823
Variable Costs[1]		
Seed Materials	1,344.91	2,689.83
Harvest Cost	1,532.38	2,978.82
Total Variable Costs	2,877.30	5,668.65
Nett Benefit ($/acre)	9,892.60	18,154.35

Gross Profit Margin

$$\frac{\text{Net Incremental Benefits}}{\text{Net Incremental Costs}} = \frac{18.154.35 - 9.892.60}{5669 - 2877} = \frac{8262}{2792} = 2.96 = 296\%$$

[1]The variable cost estimates assume that the improved planting material would cost farmers twice as much as the farmers' own planting material and that harvesting uses 30 man days/acre/10,000 lb yam at $20/man day

The trial clearly indicated that virus-tested White Lisbon yam could also perform well under subsistence conditions of the small-scale farmers. The selected small farmers on St Lucia averaged ca. 94% higher yields using the improved yam planting material. Unsurprisingly, the farmers were reportedly 'very interested in the returns they can expect from using these materials'. It is worth noting from Table 3 that the variable costs would nearly double, to a level equal to 44% of the previous year's gross revenue. It is not reported how many of the farmers felt they could afford these increased variable costs in the first year or how they perceived the risk that this investment would entail.

2.4 Developments after termination of project funding in 1984

The EDF funding ended in 1984. Since then, there has been no rigorous evaluation of the project, either by the erstwhile donor organizations or by any of the other institutions involved in the implementation phase of the project. It would seem, however, that for a number of reasons, things went rapidly and dramatically wrong after 1984.

Maintenance and planting of virus-tested material

In the first place, there were a number of problems related to the production and maintenance of the virus-free yam stocks with the result that, during 1984-86, the project ground to a halt. There was no release of improved material to Approved Growers between 1984 and 1986. Although the virus-tested stock continued to be maintained *in vitro*, some of it was lost when the air conditioning in the culture room failed in April 1985. Then in 1986, a large proportion of the different yam clones held were found to be contaminated by bacteria. Steps were taken to micropropagate only those specimens considered to be contaminant-free. Improved standards of hygiene in the laboratory were subsequently adopted.

Dr Bancroft, the plant pathologist in charge of the CARDI Tissue Culture Unit from 1986, set to work to remedy the situation. Yam plantlets derived from culture were 'weaned' onto soil and planted in the unit's screenhouse for monitoring. Mini-tubers were planted in a field plot, monitored, and replanted in 1988. In 1989, the Tissue Culture Unit successfully re-established the supply of improved yam planting material to Approved Growers for conventional mass-propagation and sale to local farmers.

Approved Growers conventionally multiplied the disease-free stock in their fields and large quantities of A, B and C grade material were sold to local farmers. During the 1985-86 season, it was estimated that 78% of the 224 ha of yam planted in Barbados were derived from virus-tested planting material.

Outbreak of anthracnose

In the 1985-86 season, at the same time as the Tissue Culture Unit was experiencing problems with contamination, a serious outbreak of the fungal disease, anthracnose (see Box 2), occurred. Anthracnose is now epidemic in Barbados and throughout the Caribbean. The yam variety originally chosen for the tissue culture work, *D. alata* c.v. White Lisbon, has proved to be especially susceptible. Consequently, yam yields have been devastated and farmers are abandoning the crop. In Barbados, the area under yam has fallen by 75%. On islands other than Barbados, anthracnose problems are often less severe because (i) farmers grow other (more resistant) species (e.g. *D. trifida*) and, (ii) farmers use stakes to support yams rather than letting them grow free on the ground.

The anthracnose problem has brought even the original findings on the yields of the new planting material into question. Recent trials (1989) compared yields of tissue culture clones from the early 1980s with some from 1987 and some 'traditional' farmers' varieties. It is certainly no longer conclusive that the tissue culture varieties out-perform farmers' yams, and the whole issue has been clouded by the anthracnose problem. Indeed, some farmers draw a direct connection between the anthracnose problem and the tissue culture process and no longer have any faith in it. There are chemical treatments available for anthracnose but farmers are reluctant to use them because (i) yam is traditionally 'only' a catch crop between sugarcane crops and has low priority for commercial sugar growers, and (ii) chemical protection is expensive and certainly prohibitive for small-scale farmers. Dr Bancroft contends that the tissue culture process should not have affected anthracnose susceptibility but concedes the possibility that the high yielding clones originally selected for tissue culture were particularly susceptible. As a result, according to Bancroft, the project is 'almost back to square one'.

Further technological solutions are being sought, this time to remedy the anthracnose problem. Research is being conducted on the transfer of anthracnose resistance to *D. alata* c.v. White Lisbon via protoplast fusion (Sinclair Mantell, Wye College). Studies funded by the UK Natural Resources Institute are attempting to reconstruct the epidemiology of the anthracnose outbreak, which is apparently also evident elsewhere in the world and could potentially be very serious - it is possible that a new fungal strain is responsible.

Recurrence of viral diseases

As if to add further misery to the disappointing last few years of the yam project, the tissue culture clones are now showing symptoms of virus infection. IBS is also recurring although apparently not seriously enough to cause commercial problems. The source of the current infections is unclear. However, concern is being expressed about some of the earlier propagation work. Two theories currently hold much credence. The first is that the core tissue culture material was freed of viruses in the original meristem culture work in the 1970s, but that poor handling by unsupervised staff (between 1984 and 1986) resulted in re-infection. The second is that even the A grade yam material still contains significant levels of infection which the original researchers, with no access to modern diagnostic methods such as ELISA (enzyme-linked immunosorbent assay) for assessing viral levels, could not detect.

General points about progress of the CARDI Unit

These technical difficulties were not the only reasons why the CARDI Unit never achieved its objective of self-funding. According to Dr Bancroft, sufficient economies of scale were simply not achievable. It became evident that Approved Growers, in seeking higher profits, were multiplying the improved varieties conventionally in their fields rather than returning to the Unit for new disease-free stock. The Unit has now also developed tissue culture procedures for cassava, sweet potato, banana, plantain, pineapple and some ornamentals including orchids. The Unit charges a 'nominal' amount for banana and ornamentals to 'non-project' customers. These activities are severely stretching the staff, which only consists of Dr Bancroft, three technicians and a secretary. The shortage of personnel arises because CARDI can only provide limited core funds for salaries and has to continue to compete for project (AID) funds which provide for equipment and recurrent costs only.

The cassava work, aimed to provide an indigenous source of animal feed, has been a technical success. However, the improved varieties are not grown because they are considered to be a 'poor man's crop' and the economics of their use are unfavourable compared with imported maize meal (from the USA). Similar problems have affected the sweet potato work.

2.5 Conclusions

The yam case shows that biotechnology projects can be achieved with moderate financial support, and can have an important impact on agricultural production. The yam project was well organized and properly managed, and contributed significantly to research capacity building in the Caribbean. Donors provided long-term support (ten years in total) which enabled the project to move from the research to the implementation phase.

The major positive point of the yam tissue culture project from the perspective of small-scale agriculture is that a 'minor' crop, like yam, which had hardly received any research attention at that time, was chosen for the study. However, it is clear that although the crop was largely grown by small-scale farmers, the research results were not automatically beneficial to those farmers. The crop, and not the use of the crop by small-scale farmers, was the target of the project. Indeed, small-scale farmers were not involved at any stage of the project cycle.

With this undifferentiated approach and the reliance on the trickle-down effect for the dissemination of the innovation, it was perhaps always likely that it would be the larger scale farmers who would benefit most from the growth of export and local processing markets. We believe, however, that the project could have readily been designed more specifically to benefit small-scale farmers. It would have been necessary to pay more attention to the farming systems of small-scale farmers and their specific problems, and to ensure that small-scale farmers were more actively involved.

The major drawback of the project was that, despite the initial increase in yam production, the subsequent and dramatic decrease meant that the innovation was virtually abandoned by farmers. The project created the technology but did not anticipate its associated problems. In that sense, the innovation was not robust. We believe that the robustness of an innovation is vital and must feature prominently in the project design and assessment.

Chapter 3

Guidelines for assessment of project proposals

With hindsight, it is easy to distinguish why projects fail or, at any rate, fail to live up to expectations. It is, however, difficult to learn from failures and to prevent them from happening again. To ensure that the limited funds available for development are well spent and that donors take more account of and are more accountable to the end-users of their services, we are working to develop a set of criteria (at present only in guideline form). Proposals for technology-based innovations can be assessed with reference to these guidelines. We believe that these criteria will help to forge a more thorough and coordinated approach not only to the funding of development projects but also to their implementation.

3.1 Criteria for admission and assessment of project proposals

It is important to distinguish at the outset between criteria for the admission of project proposals and criteria for their assessment. Virtually all funding organizations have criteria on which they base the admissibility of project proposals. These are mainly concerned with the question whether the proposals fit into the general policy framework, programme objectives and priority research areas of the funding organization. For instance, the current criteria of the Research and Technology Programme of DGIS (The Netherlands Directorate General for International Cooperation) are shown in Box 3 (DPO/OT, 1988).

Box 3: Current admission criteria of the research and technology programme of DGIS

Scope:
1. Scientific and technological research which aims to develop new methods and techniques for combating poverty and increasing self-reliance of the Third World; for research aimed at shaping and modifying Dutch development cooperation policy.
2. The research must be compatible with Dutch development cooperation policy and should, preferably, be in a field in which The Netherlands has specific expertise at an international level.

National considerations:
1. The results should be relevant to more than one developing country (replicable).
2. The project should involve cooperation with organizations in countries with which DGIS has bilateral ties.
3. Implementing organizations should make their own contributions to the project (this ensures that projects are incorporated into the organizations' normal research activities).
4. Priority is given to funding those activities (a) which are implemented by organizations in developing countries and (b) to which expertise available in The Netherlands could be transferred.

Other considerations:
1. Experts from developing countries and representatives of the potential users of research results should be involved in preparing and implementing or monitoring the research.
2. The research results or newly developed technologies should, in principle, be made available to all developing countries.
3. Research should, as far as possible, be coordinated with research elsewhere in the same field.

Obviously the admissibility of a project gives no guidance as to its quality. That a proposal could be supported is no indicator of its merit. The assessment of project proposals requires its own specific criteria.

The rationale for using specific criteria for assessment of project proposals is the assumption of a causal relationship between the following elements: project proposal ---> project activities ---> project results (outputs) ---> project objectives ---> development objectives. Application of clear and well thought out criteria will improve the identification and planning of project activities, making it more likely that the project results will be produced on schedule and be of the kind, magnitude and quality specified (UNIDO, 1984). This will increase the probability that the project and development objectives will be attained. Proposal assessment criteria, however, cannot guarantee a project's success since one can never be sure that the original project design or plan will work over time, particularly in a dynamic context. Having assessment criteria (a checklist of sorts against which proposals can be measured) is a necessary, though not sufficient, condition for the design and implementation of successful projects.

The use of assessment criteria for project proposals by organizations which fund development is a rarity. Usually the quality of project proposals is assessed by 'peer' review involving either individuals or a committee. Some funding organizations do have checklists for assessing certain aspects of proposals, for example environmental and cultural impact, but whether or not they are used apparently often depends on the officer in charge. The United Nations Industrial Development Organization (UNIDO) is one of the rare exceptions; its criteria for project design are listed in Box 4 (UNIDO, 1984). Even these criteria, however, are insufficient to assess the appropriateness of biotechnological innovations for small-scale agriculture in developing countries; they are too general and do not anticipate negative effects.

3.2   Presentation of guidelines

It is difficult to develop criteria for donor organizations that are precise enough to decide when to approve and when to reject project proposals in the field of biotechnology for small-scale farmers in developing countries. As we strive for such criteria, we have developed guidelines (Box 5). These require that proposers demonstrate that they have specifically considered certain aspects of the project. At the same time, the guidelines leave room for flexibility in the way these aspects are dealt with (depending on local tradition and culture).

More than in quantitative statements, we are interested in qualitative statements based on the best available information. The guidelines should be seen as an adjunct to the criteria for admission of project proposals. They have been developed on the basis of evaluations of earlier agricultural innovations in developing countries and recent developments in biotechnology. Each guideline may be in a different stage of development.

Box 4: UNIDO project design matrix

PROJECT LOGIC	ACHIEVEMENT INDICATORS	CRITICAL ASSUMPTIONS AND EXTERNAL FACTORS
Development programme or higher level objective(s): What is the reason for the project, the broader and/or longer range sectoral objective, problem or programme goal towards which the efforts of the project are being undertaken, who is the target group, what change, result or impact is being sought?	*Impact measures:* Can a project causal linkage be identified (in quantitative or qualitative terms) to the development or higher level objective or problem? What are the direct or indirect means of verification, i.e. how and when will UNIDO, the host government, or anyone else, know or recognize that a completed project has made the hoped-for contribution?	*Project objective to higher level objective(s):* What are the variables or complementary actions involved in accomplishing the intended impact? Which ones are critical to project relevance, i.e. impact on the higher level objective.
Project (immediate) objective and function: If the project is successfully completed, what changes or improvements could be expected in the targeted group, organization or area? Alternatively, what hypothesis or process is to be tested? What is the project specifically trying to achieve? What is the project mode of action, e.g. institution-building, direct support, etc.	*Status at end-of-project operations:* What are the conditions existing at the start of project activity, i.e. baseline data? What evidence, measures or indicators will confirm that the project's objective has been achieved? Who will undertake such a confirmation, when and how?	*Outputs to project objective:* What events, conditions or decisions outside the control of project management are necessary for the successful conversion of the outputs into the achievement of the project's immediate objective?

Project outputs/results: In relation to project purpose, duration and available resources, what are the expected or intended results of project activities which will need to be produced in order that the project objective can be achieved?

Project activities/work programme: What project activities or tasks need to be undertaken to produce each major output?

Project inputs: What goods and services, i.e. experts, training, equipment, staffing, facilities, etc,., are to be provided by (a) the government, (b) UNIDO, (c) other funding agencies or (d) other donors, to permit undertaking the necessary activities in the workplan?

Output targets and magnitude: What is the magnitude of each major output to be produced, quality or desired levels of capacity, and target dates required? If not specified, how will achievements, including progress thereon, be measured or recognized as a result of project activity?

Milestones and events: What are the milestones or major events, expressed in substantive terms, involved in the task required to produce each output and their estimated completion dates?

Budget and schedules: By each major output or event, what is the quantity, quality, and delivery date of inputs required to meet the work programme and target dates jointly agreed upon by each supplier of inputs, e.g. UNIDO and the government?

Workplan to outputs: What, if any, are the events, conditions or decisions outside the control of project management which are necessary in order for the successful performance of the activities to bring about the planned production of each project output?

Inputs to workplan: What, if any, are the events, conditions, or decisions outside the control of project management which are necessary in order for the inputs to be delivered and utilized as programmed?

In some cases, we only know that an aspect seems important for a project's success, but it is unclear yet how, or even if, it could become a criterion. With other guidelines it is already fairly clear what the eventual criterion will look like.

Box 5: Guidelines for assessment of project proposals on biotechnology for small-scale farmers in developing countries

A proposal should:
1. Demonstrate how the end-user needs have been identified, how they have been involved during the design of the project and how they will be involved during its execution.
2. Outline the anticipated economic, social, environmental and cultural impacts. Among the considerations should be:
   - The type and scale of the problem addressed;
   - The input changes implied by the innovation;
   - The output characteristics of the innovation;
   - The income-generating effects;
   - The effect on social and economic relations;
   - The effect on the robustness of the farming system.
3. Demonstrate how the generation of the proposed biotechnological innovation fits into existing rural development policy and demonstrate that it has the necessary formal and informal support.
4. Outline the institutional mechanisms envisaged both for the research and development process itself and for the dissemination of the innovation to the target group.
5. Indicate whether synergies (or antagonism) with other technological, political or economic measures exist and describe how these can be used (or circumvented).
6. Demonstrate that the proposed biotechnological innovation is both technically feasible and safe.
7. Show that the biotechnological innovation has a comparative advantage over other options.
8. Pay explicit attention to technology transfer and the building of indigenous research capacity. The mechanisms of aspects such as training and intellectual property issues should be described.
9. Give details of the organization and management of the project.
10. Stipulate realistic time-scales for completion of the project or achievement of its objectives so that the project does not fail simply through exhaustion of funds.

As well as enabling donor organizations to assess project proposals, the guidelines can be used by proposing (research) institutions to design project proposals. To show how the guidelines might work, we will give a detailed description of each guideline. When possible, we will refer to the case study on 'yam tissue culture in the Caribbean'.

1. The proposal should explicitly demonstrate how the end-user needs have been identified, how they have been involved during the design and how they will be involved during its execution.

As standard, any proposal indicates the intended beneficiaries. However, we consider this insufficient. Rather we think that donors should insist on knowing explicitly how the target group has been identified. Furthermore, they should seek evidence that the proposers have identified a 'real' need of this group. The reason for this stringent approach is that, if technological innovations are to be adopted, the participation of the farmers and the conditions under which they operate are key variables. They will determine the success of the final application (Bunders, 1988; Elz, 1984). A strong proposal, therefore, would demonstrate a high degree of farmer participation in the design of the project. Mechanisms for this include the 'interactive bottom-up approach' or participatory technology development approaches (see Part Two). If the proposing organizations do not themselves have good working links with end-user groups or expertise in participatory project identification procedures, they should collaborate with an organization that does.

End-user participation is an essential component of any innovative technology-development process (Kaimowitz and Merrill-Sands, 1989; World Bank, 1990). Therefore, the proposers should outline how farmers participate not only in problem identification and project formulation, but also during the different stages of project implementation. Farmers' participation should include more than just the use of their land and services. Also it is not enough just to consult them about their problems. Farmers should be involved in the development of solutions based on their indigenous technical knowledge (see Box 6; Biggs, 1989).

In our yam example, the original research addressed a problem relevant to both large- and small-scale farmers, and to a substantial proportion of consumers. The target group of the project was the yam farmer in general and, not explicitly, the small-scale farmer. Small-scale farmers were, by implication, the major target group: the great majority of yam production in the Caribbean is by small-scale farmers on small plots (of less than 0.5 ha). Furthermore, the variety selected for the initial research, *Dioscorea alata* c.v. White Lisbon, was widely grown by small-scale farmers. However, the small-scale farmer was never identified explicitly as the target end-user. During the execution of the project, the emphasis shifted towards commercial yam production systems, and thus towards the larger yam producers. Indeed, this shift is illustrated in the Final Report 1980-84 which states that 'the project in fact functioned as a catalyst to address some of the(se) problems of the yam industry' (CARDI, 1984).

Box 6: Modes of farmer participation

ISNAR (International Service for National Agricultural Research) distinguishes four modes of farmer participation in research:

Mode	Objective
Contractual	Scientists contract with farmer to provide land or services
Consultative	Scientists consult farmers about their problems and then develop solutions
Collaborative	Scientists and farmers collaborate as partners in the research process
Collegiate	Scientists work to strengthen farmers' informal research and development systems in rural areas

---

Participation of farmers during the project was limited to (i) provision of land for on-farm research; and (ii) attendance of farmer field days between 1980 and 1984. The mode of farmer participation can, therefore, be considered largely contractual. For instance, farmers were excluded in the selection of yam clones. We would draw the conclusion that the farmers and the conditions under which they produce were not key variables for the yam project. If the conditions under which these farmers operate had been taken into account, it is possible that more disease-resistant varieties would have been preferentially selected, and that the projections on the potential for self-funding of the Unit would have been amended.

2. The proposal should outline the anticipated economic, social, environmental and cultural impacts.

The introduction of technological innovations can have profound effects on the production capacity of the farming system. Small-scale farming systems depend on the delicate interrelationship between four major elements which renders them highly vulnerable. The four are (i) biotic elements (plants, animals, micro-organisms), (ii) abiotic elements (water, temperature, soil texture and structure, nutrients, etc.), (iii) other 'internal' elements (draught power, land, processing facilities, agricultural practices, availability of labour, etc.), and (iv) 'external' elements (fertilizers, seeds, market changes, institutional linkages, etc.) (Bunders *et al.*, 1990). Farming systems are rarely static; they change. Proposals for innovations, therefore, should consider this interrelated and 'dynamic' context.

Although there is currently no systematic method for anticipating economic, social, environmental and cultural impacts of innovations, a number of aspects can be distinguished:

1. The type and scale of the problem addressed by the innovation.
2. The requirements for application of the innovation; the inputs needed and the characteristics of the outputs.
3. The potential effects of applying the innovation on income-generation, social and economic relations, and the robustness of the farming system.

These aspects will be described below in more detail and illustrated with examples from the yam project.

The type and scale of the problem addressed

A thorough description should be given of the problem addressed by the proposed innovation. What is its nature? In which geographical area is the problem most acute? What are its consequences for the target group?

In the yam example, the project proposals did give a fairly extensive description of the type and scale of the problem to be addressed: the disease was identified, the extent of infection was documented and the scale and variety of the inputs and outputs were detailed.

The input changes implied by the innovation

Small-scale farmers generally have limited access to inputs (Bunders, 1988; World Bank, 1990). Thus the most appropriate biotechnological innovations will usually be those which neither require significant inputs nor significant changes in inputs. There are two different types of input: internal and external (Bunders *et al.*, 1990). The former are inputs which draw on existing resources such as the means of production (land, draught animals, capital, technological equipment and labour), the land use pattern (the agricultural calendar of the farming system), the distribution of labour, and the socio-cultural patterns. The latter includes capital-intensive inputs, credit facilities and education services.

None of the input requirements for application of the yam tissue culture project was explicitly defined in any of the project proposals. No mention was made that farmers would have to buy seeds and adopt new cropping and harvesting techniques. It subsequently transpired that the variable costs (seed and harvesting) nearly doubled when the new technology was adopted. When it was proposed that mechanization and mono-cropping should be adopted by small-scale farmers, this would require prohibitively large investments. Had these factors been considered, it would have been clear that small-scale producers could only benefit from the disease-free varieties if they could meet the increased variable costs. It was not reported how many of the small-scale farmers felt they could afford these increased variable costs in the first year or how they perceived the risk that this investment would entail. Otherwise, the proposal might have been amended to provide measures like credit facilities for small-scale farmers.

The output characteristics of the innovation

The output of the innovation should meet the demands of the target group. That demand needs to be defined as precisely as possible. This is especially true where the innovation affects products which have a multifunctional use. Appropriate innovations will take this aspect into account.

Three different kinds of characteristics of the output should be considered explicitly (Bunders *et al.*, 1990): nutritive value, other uses of the product, and its quality. Biotechnological innovations may change the nutritional characteristics of the output in such a way that the amount of food consumed by small-scale farmers' households is changed.

Alternatively, or concomitantly, an innovation may change characteristics in a way that affects its use for other productive or remunerative purposes within the farming system. Small-scale farmers are usually involved in a variety of complementary and interchangeable productive on- and off-farm activities. It may be insufficient, therefore, simply to consider the nutritional value of a crop. Whether or not an innovation is adopted may depend on whether the new output can be used in existing productive or remunerative farm activities or, indeed, to create new ones (Bunders, 1988).

A biotechnological innovation may change the quality of the output for marketing. For example, biotechnology could improve the quality (and quantity) of the output through new or improved on- or off-farm processing techniques (fermentation). Furthermore, output characteristics can be changed by the introduction of biotechnologically improved crops.

The income-generating effects

In order to realize income, the outputs of any innovation have to be exchanged either for cash or for other goods. In assessing the potential income-generating effect of biotechnological innovations, therefore, it is important to consider not only whether there is a market for agricultural crops and a mechanism for reaching that market, but also what policy measures are needed to support the small-scale agricultural sector. Where there is no market for certain agricultural products, or where the markets are highly unstable (e.g. as a result of competing imports and unguaranteed prices) the income-generating potential of the biotechnological innovation may be severely limited. The potential will be even more limited if transport and storage facilities (road systems, harbours, availability of transport and field depots) are poor. Where this is the case, it will be totally unrealistic not to point out in the proposal the need for additional measures and programmes. A proposal should, then, address issues related to the establishment or improvement of regional/local markets.

In the yam tissue culture project, the income-generating prospects looked good. Adoption of the innovation would enhance income either in cash or kind for farmers who produced for the market. There was already an extensive local and export market for yam of good quality. On the larger scale, the restoration of the yam export trade and the reduction in food import bills (regional food self-sufficiency is low) would have had positive implications for overall regional economic prosperity.

The effect of the innovation on social and economic relations

A proposal should consider the scope of the innovation for influencing the social and economic circumstances of the target group. The social considerations have many dimensions. An innovation could have an influence on the dynamic within the farm holding itself - the balance between the roles and goals of the men, the women and the children there. It is likely to have differential effects on different categories of farmers: large-scale, small-scale, or landless farmers. Probably, too, it will affect the relative circumstances of groups in the rural chain - farmers, landless labourers, consumers, traders, and officials. Finally on the social level, it may invoke effects at national level, altering the relative positions of rural and urban populations.

Economically, the innovation could influence the dynamics of production, creating shifts between subsistence and market-oriented production, between production for local and export markets, and between informal and formal production.

The effect of the innovation on the robustness of the farming system

Determining the robustness of an innovation is a matter of assessing its potential influence on the stability of the output and on the sustainability of the farming system (Bunders *et al.*, 1990). An innovation should not adversely affect the stability of the output; the output should be at least as resilient as the original output to the normal stresses and fluctuations to which it is exposed. Pests and diseases, drought or frost periods, and the timely availability of inputs may perturb outputs, but any innovation should not increase that perturbation.

At the same time, an innovation should not result in an exhaustion of the basic resources of farming - the biotic and abiotic resources like gene pools of plant and animal species, soils and water, the economic resources like capital and labour, and the socio-cultural resources like organization of labour, cooperation and mutual help practices, and traditional agricultural know-how.

Currently, only a few types of small-scale farming systems in developing countries are stable and sustainable. Most systems are in the midst of changes that have been induced by factors such as land shortage due to population pressure, or the earlier introduction of inappropriate technology (Wolf, 1986). Moreover, the effect of a reduction of the basic resources due to a new technology is more pervasive in small-scale farming systems than in any other farming systems because small-scale farmers usually live on marginal soils and have limited access to inputs.

Robustness of an innovation should be examined extremely carefully. It can sometimes be illusory. For example, biotechnologically-acquired, disease-resistant crops may well increase the stability of the output in the short term. Then, with the disease thwarted, productivity may be increased by the adoption of apparently more efficient crops. However, in the longer term the stability of the output may decrease. If the disease resistance is subsequently lost - 'breakthrough' occurs when the disease-causing organism finds a way round the plant's resistance - then the crop becomes highly vulnerable to disease. With the introduction of the innovation knowledge of ways of reducing diseases by traditional cropping patterns may be lost. Even if that knowledge is retained, its reimplementation could take many years, time which subsistence farmers do not have. Thus an apparently robust innovation can ultimately reduce the sustainability of the farming system.

In the yam tissue culture project, there was no information in any of the project proposals with respect to robustness. That the project was far from robust became evident upon subsequent evaluation. The project did result in the virtual eradication of disease symptoms from a popularly grown variety of yam, *D. alata* c.v. White Lisbon. At the same time, the tissue cultured clones dramatically improved yield and quality on farms. This was definitely a contribution to the stability of the output. Several years later, however, the region suffered a severe outbreak of the fungal disease, anthracnose. The same yam variety, *D. alata* c.v. White Lisbon, fared the worst; yields dropped disastrously and many farmers abandoned the crop (some blaming the tissue culture work for their problems). Also several viral diseases recurred. The early apparent 'success' of the project appeared to be unstable. It could be argued that the impact of the diseases would have been limited had pathogen resistance (and not merely yield and quality) been a key criterion for selection of the original clones.

3. The proposal should demonstrate how the generation of the proposed biotechnological innovation fits into the existing rural development policy of the recipient government, and demonstrate that it has the necessary formal and informal support.

The proposers of an agricultural innovation should take into account the prevailing rural development policy and state how the project is plausible within the context of that policy. If a developing country's government has no clear political intentions towards small-scale farmers, it would be difficult to integrate any project into current research and development programmes. Then, additional efforts, such as the mobilization of the necessary formal and/or informal support, would be essential to assure effective execution and continuation of the project.

The extent of support that the project can expect from relevant persons and groups (including scientists, politicians, marketing agents, and women, consumers' and farmers' organizations) should be indicated in the proposal. Within a practical time-frame for project preparation, it would usually be impossible to assess thoroughly the support that might be obtained from all relevant parties. However, there are methods for assessing the extent of support and these can often be executed with relatively limited resources and in a short period of time; an example is the 'interactive bottom-up approach' (see Part Two). Whatever mechanism is used to assess or solicit support, we would recommend direct collaboration with one or more indigenous organizations which have the requisite links to various parties.

In our project example, research to improve the yield, quality and overall output of yams was perceived as achieving both agricultural policy objectives (increased export revenues, increased regional food self-sufficiency) and science and technology policy objectives. The project received broad support from the government and from relevant scientific and agricultural institutions: collaboration was fostered between the participating institutions (CARDI, Barbados Ministry of Agriculture and The Barbados Sugar Producers Association). However, the proposers did not attempt to win the support of other groups.

4. The proposal should outline the institutional mechanisms envisaged both for the research and development process itself and for the dissemination of the innovation to the target group.

An industrial organization would be extremely unlikely to commit substantial funds to research, development and commercialization of a new product unless it was confident that there were effective mechanisms for translating R&D into marketable products. Similarly, a development funding organization should not provide funds for innovations aimed at small-scale farmers unless it is convinced that mechanisms are or will be in place to ensure effective delivery. The nature of the delivery mechanisms is less important than its effectiveness: they could result from the proposers' own independent efforts or through collaboration with rural extension agencies, NGOs or other organizations which have the requisite skills and resources and a proven track record of effective interactions (as intermediaries) with the relevant target group of farmers and with researchers. It is important that a proposal provides an inventory of available dissemination mechanisms in a specific area. The proposal should consider whether and how, in the absence of specific action, dissemination of the innovation can take place. In most developing countries, organizations for the dissemination of technological innovations exist; their adequacy in reaching small-scale farmers in marginal areas is, however, questionable and may require considerable improvement.

In the yam tissue culture project, the Caribbean Agricultural Research and Development Institute (CARDI) was responsible for the R&D process, with its most immediate objective the multiplication of elite yam planting stock. The dissemination of the innovation involved a number of mechanisms. One of the longer term aims was to establish a tissue culture unit in Barbados to keep up the production and maintenance of 'nuclear' stock and the propagation of 'elite' planting material. The innovative 'Approved Growers' scheme was devised to effect continuous distribution to farmers in Barbados and other countries in the region. The selected group of 'Approved Growers' in Barbados would purchase 'elite' planting material from the tissue culture unit and re-sell their output of yams to both commercial and subsistence growers as 'Certified Disease-Free' seed tubers. Small-scale farmers were indeed reached through this mechanism although to what extent is unclear.

5. The proposal should indicate whether synergies (or antagonism) with other technological, political or economic measures exist and describe how these can be used (or circumvented).

The introduction of a biotechnological innovation could be said to be synergistic if it facilitates or is facilitated by the preparation/implementation of another innovation. Any project should try to make the best use of the possibility of synergies. Project proposers, therefore, should investigate if other, possibly complementary, projects are running or have already been completed. The introduction of a biotechnological innovation with strong synergies is much easier than those which run into antagonism. A common form of antagonism, for instance, results from trade liberalization. This creates free but unequal competition in the internal market between imported and locally produced goods and in so doing hampers the introduction of a biotechnological innovation designed to increase local production of a particular product.

6. The proposal should demonstrate that the proposed biotechnological innovation is both technically feasible and safe.

Technical feasibility can be described in terms of the state-of-the-art of the proposed technology, the time-frames for R&D, the budget, the scientific capability of the research institutes, and other conventional criteria for the assessment of scientific proposals. The sort of questions that might be asked include the following: Will the design and method of the experiment or study efficiently and effectively allow conclusions to be reached? What is the probability of technological success? What is the track record of the proposers? Will specific regulations (like good manufacturing practices, rules for working with genetically modified organisms in the laboratory) have to be applied?

It is not sufficient, however, to consider the performance of a project in the laboratory or research station. Agricultural innovations for small-scale farmers must succeed at the farm level and under the prevalent conditions and practices there. A research and development process is only useful if it can generate innovations which are widely adopted and which produce stable outputs and sustainable development. It would be obviously impractical to undertake biotechnology research on farms. Ultimately, however, the farmer is the final arbiter of 'technical success'. Therefore, in making biotechnological R&D proposals, the problems should first be diagnosed with farmers, and during the research, outputs should be adapted according to an iterative process which involves farmers.

When the yam tissue culture project started in 1973, it was unclear whether the proposed innovation would become technically feasible. After 6 years of research, the researchers clearly felt that the production of disease-free yam tubers by tissue culture was technically feasible. The question of safety was not broached and, perhaps, did not arise. However, it could be argued that adherence throughout the project to good manufacturing practice (ostensibly a safety measure) would have avoided some of the problems encountered later with bacterial infection of the tissue culture clones.

7. The proposal should show that the biotechnological innovation has a comparative advantage over other existing options.

In a drive for the most efficient use of development funds and research efforts, proposals in biotechnology should demonstrate that the use of biotechnology has a comparative advantage over other options. Four distinct aspects should be considered: (i) the anticipated economic, social, environmental and cultural impacts (see Guideline 2); (ii) the technical feasibility and safety (see Guideline 6); (iii) the problem-solving capacity; (iv) the cost-effective delivery of solutions to problems.

The process of weighing the relative advantages of a laboratory-based biotechnology research process and a 'low tech' on-farm approach impinges on the classical debates in development - 'centralized versus decentralized' and 'formal versus informal' R&D. Consider, for example, the various ways of reducing crop losses due to disease. One could envisage that the laboratory-based solution might be to insert a gene for disease resistance into a relevant crop. However, with appropriate assistance and support, farmers might achieve precisely the same objective with a new system of crop rotation and/or by planting disease-resistant material used by farmers elsewhere. This decentralized or informal approach has the advantage that it promotes and strengthens largely indigenous processes. It is more likely to result in a sustainable development. The proponents of 'high-tech' laboratory-based research, on the other hand, may feel they are 'pushing back the frontiers' of science. This, however, is not always and automatically an advantage to a developing country's national research system, even though it may boost the morale of individual scientists.

The proposers of the yam tissue culture project did address the question of comparative advantage. They argued that tissue culture techniques would have a number of advantages over other options. For instance, they estimated that, over a two year period, the micropropagation of segments from a single 100 g tuber would produce more than 300,000 kg of disease-free 'seed' tubers for planting as opposed to around 25 kg by 'traditional' propagation methods (Mantell *et al.*, 1979). They pointed out that unlike conventionally propagated tubers, those produced by tissue culture were of consistent high quality and shape, and free of brown spot or rotting symptoms. On cost grounds, they argued that biotechnology was a way in which thousands of elite yam clones could be prepared and maintained in a small space; traditional methods demanded extensive and costly greenhouses and/or field collections. They argued, too, that the tissue culture method not only provided a better solution but also one that could be extended to treat the same problem elsewhere. Maintaining 'nuclear' disease-free stock under aseptic conditions *in vitro* meant that new, elite planting material was always available to combat reinfection in the field. That some stock could also be transferred from country to country, or region to region, without the risk of introducing pathogenic soil micro-organisms, was clearly an advantage over other existing methods.

8. The proposal should pay explicit attention to technology transfer and to the building of indigenous research capacity. The mechanisms of aspects such as training and intellectual property issues should be described.

Stimulating the involvement of scientists in developing countries is a central requirement in achieving a sustainable contribution of national research to small-scale agriculture. Building indigenous research capacity will enhance the self-reliance of developing countries - one of the core objectives of the research and technology programmes of many donors. But in order to strengthen local research capacity, a project will have to pay explicit attention to technology transfer mechanisms and local research capacity building. It will be important for the proposal to identify not only which counterpart research organization in the developing country will be involved but also to describe its track record. It may become clear that certain aspects of this counterpart research organization need strengthening: the proposal should state what these are and should suggest remedies.

Two issues require particular attention. The first arises when the biotechnology research takes place completely outside the country concerned, when key skills and resources cannot be 'captured' by the indigenous research system. Those advocating that the biotechnology research should take place in an industrialized country, ignore the advantages of local research capacity building. Instead, they point to the economies of scale in research in the industrialized country.

The involvement of a commercial organization in the project, either as a participant or as a co-funder, can create another problem: the issue of intellectual property rights and access to research results. While any extended dispute over such matters could create financial setbacks for the company, the delays it would cause in the application of an innovation in the developing country might be disastrous.

When such issues are expected to arise during a project, the proposal should indicate what specific arrangements will be made to prevent problems. For example, if it is inevitable that the research programme is executed mainly outside the country concerned, the project might 'compensate' the indigenous research system by making a commitment to the active participation and training of national scientists. Similarly, any commercialization process should acknowledge the need of the 'recipient' national agricultural research system to have access to the information and specific technologies generated by the project on reasonable and negotiable terms. Specific contributions of a collaborating national agricultural research system should be recognized as should the specific need of the recipient nation and its people. Both aspects could be addressed by agreements (made in advance of any development) to grant licences for specified uses to the participating parties, for example. There will be some cases where indigenous genetic resources, including those already significantly 'improved' by farmers and researchers in the 'recipient' country, are commercialized. Under these circumstances, the payment of appropriate royalties would seem entirely reasonable. The proposal clearly should address this issue.

In the yam tissue culture project, technology transfer and capacity building was not only considered, it was in fact one of the objectives of the project. The project involved active participation of local researchers and resulted in the transfer of key biotechnology skills and resources (including a custom-built tissue culture laboratory in Barbados) to the region. Caribbean researchers were involved at all stages of the research process (with support from established tissue culture experts). The new tissue culture and propagation unit had the opportunity to target more small-scale farmer crops in future: work on cassava, for instance, was initiated at an early stage. However, since no specific small-scale farmer orientation was taken in this project, the mechanisms for targeting and reaching small-scale farmers were not introduced. Problems related to specific training requirements and intellectual property rights were not expected to arise (and did not arise).

9. The proposal should give details of the organization and management of the project.

Any research and development project requires an effective organization and proper management to convert the resources allocated to the project into results within an agreed time-frame. To apply the sort of criteria that we advocate, project managers will be required to (i) process multidisciplinary data, including technical, economic, social, cultural and environmental information on the proposed technology, and (ii) adopt an end-user perspective in decision-making, even when the research takes place at a site geographically remote from its 'clients'. The adoption of a joint-venture framework creates its own problems. There will be organizational complications in coordinating the activities of the participating groups and, quite possibly, problems due to the different status or perceived status of the collaborating organizations (Kaimowitz and Merrill-Sands, 1989).

In this regard, the project proposal needs to demonstrate that those involved in its management have sufficient breadth of expertise and flexibility to handle effectively the complex issues involved. It needs to indicate the division of responsibilities at each stage of the technology identification, generation and dissemination processes. The proposal should also indicate that effective channels of communication with the intended 'clients' of the research will be set up or are in place: e.g. rural extension services or farmers' organizations. Finally, the proposal should indicate the ways in which research managers are ultimately accountable to these 'clients'.

Special attention should be paid to monitoring the R&D process. Only with a frequent and continuous activity in which progress is assessed relative to set targets can delays in, or divergence from, planned patterns be spotted and corrective action taken (Elz, 1984). Many influences on a project (e.g. government policies, legislation, price fluctuations) will remain outside the project management's control. Nevertheless, these aspects, too, need to be monitored so that corrective measures can be taken during the implementation of the project (UNIDO, 1984). The project proposal, therefore, should indicate how monitoring will take place and how frequently progress reports will be issued. Progress reports should be both retrospective and prospective: they should develop backward links to the programming of operations and forward links to operation control and evaluation (Chambers, 1985).

To provide a suitable multidisciplinary input, there may be a case for appointing a steering or advisory committee made up of individuals representing the key groups involved in and affected by the project; researchers, policy-makers, farmers, consumers and relevant technical 'experts' (economists, agronomists, etc.). The committee would require executive authority lest it fails through redundancy. It should be composed of individuals with a positive commitment to the project's success (i.e. those with an incentive to see the project succeed). This will help to avoid the research being rendered less effective by conflicts arising out of radically differing perspectives.

The proposals for the yam tissue culture project paid explicit attention to organization and management. In practice, too, the project was well organized and properly managed. Planning at all stages was thorough, carefully focused and based on empirical evidence. The objectives of the project were never over-ambitious, and were allowed to evolve over time as each stage was successfully completed. Moreover, key individuals put in a great deal of work over a sustained period.

Continuous monitoring of the project was carried out by project staff. Annual and terminal reports were produced and submitted to the Caribbean Development Bank as outlined in the project proposal. However, farmer participatory methods were not used during monitoring and evaluation.

An advisory committee was established, made up of representatives of the participating institutions (regional research network, regional bank, aid funder, local ministry of agriculture and local organization of sugar producers); chairmanship rotated between the members. Each of the represented organizations also contributed resources to the project such as financial support, land, manpower, or administrative facilities.

10. The proposal should stipulate realistic time-scales for completion of the project or achievement of its objectives so that the project does not fail simply through exhaustion of funds.

Sufficient long-term financial commitment is needed. Biotechnology R&D processes which aim to provide sustainable solutions for problems in the small-scale farming sector in developing countries are likely to have extensive time-frames. This is particularly so when research also has to contribute to indigenous research capacity building. However, donor agencies will usually finance a project only for three or four years. When longer periods of R&D and implementation are necessary, one solution may be to divide the period for financing into separate three- or four-year stages, each with clear attainment objectives. This gives the donor the opportunity to terminate financial support if progress of the project is unsatisfactory. In principle, however, satisfactory progress should ensure funding for the next stage.

One way of ensuring that the recurrent costs of a technological development project are met, is to build in a self-financing capability right from the start. The small-scale farming sector of developing countries is usually considered a high risk commercial investment. However, a rurally based self-financing biotechnology 'unit or project' need not be as far-fetched a notion as it sounds, especially if some of its costs are defrayed by state support (e.g. initial capital costs, fiscal incentives).

With respect to the yam tissue culture project, donor funds were available for a total of ten years. This meant that there was continuity in planning and the project could build on early successes. However, ultimately, the CARDI Unit did not achieve its objective of becoming self-funding. Once AID funds for yam project core costs were no longer available, funding and support for the new Unit was inadequate. Subsequently, the Unit appears to have been overstretched by focusing on diverse (AID-funded) project work and the day-to-day work of the yam project suffered.

Chapter 4

Practical implications

Biotechnology innovations can have considerable impact. This justifies that special attention is paid to project proposals. Ultimately, we believe that a set of stringent criteria can and should be applied both in the preparation and assessment of biotechnology project proposals for small-scale farmers. Our proposed guidelines (Chapter 3) are not yet criteria. Eventually, one would hope to be able to ease the burden on those assessing projects by providing questions which could be answered 'yes' or 'no'. This, however, will require extensive evaluation of many projects and proposals in biotechnology and elsewhere. In anticipation of the formulation of suitable criteria, we want to pose two key questions: 'Is the application of criteria realistic?' and 'Are there dilemmas in applying criteria?'

4.1 How realistic will it be to apply the criteria?

It is clear to us that if biotechnology research projects are ever to make a contribution to small-scale farmers in developing countries, then the proposals must be appropriate and feasible. The best way to make them appropriate and feasible is to apply criteria. We believe, therefore, that the application of criteria for the assessment of project proposals is a necessity and is realistic. Stringent criteria are frequently applied by industry to investments in innovation and product development. Mindful of its own strengths and weaknesses, aware of its past successes and failures and of those of its competitors, and sensitive (above all) to the need to deploy its resources cost-effectively, industry and those investing in industry apply criteria rigidly. Many of industry's criteria are applied automatically and unnoticed. If they were not applied, commercial disaster would be likely to strike.

The guidelines we suggest have clear parallels in some of the ways in which industry operates. Some of the industrial analogies are mentioned in parentheses. In essence, our guidelines are as follows. The target group (consumer) for the proposed innovation (product) should be defined and evidence should be provided that a genuine need (demand) of the target group is identified. One needs to ensure that the innovation will perform properly under prevailing conditions and management procedures (conforms to manufacturing practices), and that the robustness of the farming system is not negatively affected by the innovation. Furthermore, the design of a project needs to take into account trends such as population growth and the dynamics of social and economic relations (market projections). Proposers need to ensure that the research process maintains close and on-going links with, and is ultimately accountable to, its consumers/clients (market- rather than technology-led development). There must be effective channels to distribute the new technology (distribution mechanisms). The proposers involved need to demonstrate an awareness of alternative approaches (competition analysis) and that the proposed project is the best way of obtaining the stated objective. They must be convinced that the innovation will actually be broadly welcomed by various different groups who will be affected by it (assessing market sentiment). The project must be managed in such a way that complex multidisciplinary data can be processed and the right decisions made (control in-house development). Feedback mechanisms which refer to specified achievements (product development milestones) must be instituted to guide the project.

In the context of developments for small-scale farmers, such criteria provide a useful checklist for obtaining information needed by an experienced review committee. Application of such criteria would improve the efficiency and effectiveness of projects. However, more time and money will have to be spent in the preparatory phase and this would require institutional adjustments from both proposing and funding organizations.

Implications for proposing organizations

When confronted with a wide-ranging list of considerations which go far beyond the ground with which they are familiar, many scientists may feel dismayed. Some of the criteria they may view as obstructionist because they deflate their good intentions or irrelevant because they are not technologically based. Many proposing organizations will simply not have all the skills and resources necessary to meet or even address the criteria. The very existence of extensive criteria may, therefore, discourage scientists from preparing proposals.

It should not be assumed that we are suggesting that would-be proposers need to become development experts overnight. They do need, however, to become flexible and inventive in gathering relevant information. They may need to learn, at least, how to communicate with those involved in development. Proposing organizations could, for example, consider collaborating in a 'joint venture' with one or more groups with complementary skills and resources necessary to achieve specific objectives. This, too, is common in industry. Such an arrangement could help in (i) the identification of appropriate biotechnological innovations for small-scale farmers, (ii) the implementation of an applied/adaptive research stage at farm level, and (iii) the widespread extension of the finished 'product'. Jointly, these groups could submit proposals which were far more comprehensive (going beyond mere technology), far more realistic and far more likely to attract development funding.

Implications for funding organizations

We have shown that a great variety of factors need to be considered in selecting projects which identify and effectively address the complex needs of small-scale farmers. In its selection procedure, therefore, the funding organization requires a multidisciplinary perspective, which would probably go beyond the capability of any individual project officer. We would, therefore, propose that a special review committee should be appointed by any institution which funds biotechnology R&D for small-scale farmers in developing countries. The committee should have the requisite expertise to consider not only technological but also social, economic, environmental, cultural and project operational aspects.

Furthermore, in this field, evaluation should become a standard procedure for all projects, instead of a decision which is up to the funding organization. Evaluation is an important source of feedback to enable priorities to be more precisely set and programmes to be better planned (Elz, 1984; UNIDO, 1984). It is not only a comparison between what was planned and what was achieved but also an analysis of unanticipated effects.

Evaluation is generally considered a difficult process which faces serious problems of method, of experience and motivation among evaluators, of organizational and political relationships, and of choices of resource use (Elz, 1984). It needs, therefore, to be planned and managed very carefully. Evaluation can best be done by people who are not directly involved in the project. A visiting committee with representatives of a broad spectrum of interested parties would be well placed to collate and analyse the necessary information. It is, however, essential that the opinion of the intended beneficiaries, or end-users, will be included in the final report. Some of the mechanisms advocated above could be useful in this context. For example, the evaluation of the impact of a biotechnological innovation could be undertaken by a 'joint-venture' partner with the requisite end-user linkages. Disseminating the results of evaluations widely is important even, or perhaps particularly, in the case of failure. A project that fails quietly is a project waiting to be repeated and to fail again.

With the use of criteria, the preparatory phase of a project will become much more important. The information demanded implies substantial activity before a project proposal can be submitted and this will increase the costs of the preparatory phase significantly. Currently, most funding organizations require proposers to pay these costs themselves. Some arrangements should be made to ensure that at least part of these costs will be covered by a funding organization.

4.2   Are there dilemmas in applying the criteria?

There are some project proposals which would be discarded when the criteria are strictly applied, but for which one might want to make an exception. These exceptions can only be made if certain additional conditions are fulfilled or possible adverse impacts are estimated. In industry, high-risk projects are usually included in a special portfolio which allows the risk to be spread within the organization. Such an approach is undesirable in the context of small-scale agriculture; researchers, policy-makers, project officers and project managers should never decide on the risks small-scale farmers should take; 'those on the edge of survival cannot afford to gamble' (World Bank, 1990). We suggest instead another approach - one which is also used by governments and industries - the 'case-by-case, step-by-step' approach.

We have identified two categories of high-risk projects: (i) those which involve the release of organisms containing recombinant DNA; and (ii) those which offer great potential to improve the situation of small-scale farmers but which require substantial changes in current farming practices.

The introduction of any novel organism into a fragile environment may disturb the ecological system. The environmental release of genetically modified organisms (GMOs) creates particular concern, although ecologists and biotechnologists disagree on the risks that are associated with it. Experiments with GMOs in countries such as The Netherlands and Denmark must be licensed and are subjected to an extensive risk assessment on a case-by-case, step-by-step basis. However, there are no comparable regulations in developing countries. To prevent developing countries from becoming test grounds for experiments with GMOs, a minimal requirement would be to apply the regulations of the donor country. If the donor was an international organization, it could consider following the recommendations made by the Organization for Economic Cooperation and Development (OECD) in its report *Recombinant DNA Safety Considerations* (OECD, 1986). Without guarantees on safety, funding organizations should not finance a project.

It is generally held that projects which could significantly improve the situation of small-scale farmers are those that would require substantial changes in current farming practices. These, however, also involve higher risks for small-scale farmers (see Guideline 2, Chapter 3). In such cases, we would propose a case-by-case, step-by-step approach similar to the one advocated for industrial management of innovations involving high financial risks. In order to anticipate negative effects, all problems should be extensively assessed in close collaboration with the intended beneficiaries at each successive step of the project. In this case, if thorough monitoring and evaluation using participatory procedures cannot be guaranteed, then the funding organization should not finance the project.

PART TWO

AN INTERACTIVE BOTTOM-UP APPROACH IN AGRICULTURAL RESEARCH

Joske F.G. Bunders
Annelies Stolp*
Jacqueline E.W. Broerse

Annelies Stolp, Department of Biology and Society, Vrije Universiteit Amsterdam

Chapter 5

Different approaches to technology development for Third World agriculture

5.1   Introduction

The guidelines we presented in Chapter 3 constitute a valuable instrument for formulating and assessing project proposals in the field of biotechnology for sustainable agricultural development. In itself, however, such a project-by-project approach is not likely to lead to a systematic utilization of the potential of biotechnology. At most, the guidelines influence organizations which already design and/or fund biotechnology projects and programmes for small-scale farmers in developing countries.

To ensure a structural focus on small-scale farmers' needs, governments, research institutes, donor organizations and other providers of support should incorporate the research priorities of small-scale farmers into their general (biotechnology) policies. The lack of exchange of information and perceived conflicts of interest between these actors, however, impedes any consensus. In Part Two, we will argue that an approach which enhances information exchange is needed. It must identify the specific needs and problems of small-scale agriculture and must prioritize biotechnological research accordingly. Such an approach should enhance the incorporation of these priorities in biotechnology research agendas. In the development of this new approach, much can be learned from previous analyses of various other approaches to agricultural technology development in developing countries and implementation strategies for innovative projects in general.

In this chapter we will analyse past and present approaches used in agricultural research for developing countries. Three major approaches to technology development in agriculture can be distinguished: top-down transfer of technology, farming systems research, and participatory technology development. Each approach has its merits and deficiencies. 'Topdown technology transfer', for instance, frequently fails because it neglects the context in which technology at the small-scale farm level must operate. It's merits are also clear; an impressive increase in food production in various areas. 'Farming systems research' and 'participatory technology development' methods demonstrate that it is indeed possible to formulate projects based on the demands of small-scale farmers. Where these approaches have all failed so far, however, is in influencing research agendas.

In this respect, much can be learned from analyses of innovative processes. In Chapter 6, we will examine the ways in which entrepreneurs have succeeded in implementing innovations within their organization. We will also look at how industry influences public research, and at how other less powerful groups have successfully written their priorities on public research agendas.

These analyses lead us to formulate a model for farmer-led innovation processes: the interactive bottom-up approach (Chapter 7).

5.2   Top-down transfer of technology

Throughout much of agricultural research and development, both at an international and a national level, top-down technology transfer approaches are in force. Top-down technology transfer presumes a predominantly one-way, 'top-down' flow of information and technology: from 'centres of excellence' via extension agencies to farmers in general. This approach is built upon the premise that universities and (centralized) research institutes are the sole sources of new technologies. Researchers develop technologies in laboratories and experimental stations. They then attempt to transfer them to would-be clients (Chambers and Ghildyal, 1985). Those farmers with the best contact with researchers and extension workers, or those who are naturally progressive and/or innovative, will adopt the technological innovation first; a 'trickle-down' effect is assumed to disseminate the innovation to other farmers. Another assumption is that scientists adequately understand the prevailing constraints to crop production, and can make the correct decisions to generate useful technologies (Horton and Prain, 1989).

Encouragement for these assumptions has come from the success of this approach in industry and in resource-intensive agriculture. However, 'trickle-down' to small-scale farmers is not much in evidence: only occasionally do small-scale farmers adopt the innovations successfully. Noting the poor adoption rates in this sector, researchers attributed the problem either to the ineffectiveness of the dissemination mechanisms for technology and inputs (e.g. via extension, credit and input supplies), or to the inefficiency and 'ignorance' of small-scale producers (Chambers and Ghildyal, 1985).

During the 1960s and 1970s, however, it became clear that the inappropriateness of the technology was the root cause of non-adoption; innovations designed for well-regulated, 'industrial' agriculture were ill-suited to small-scale farms. It was recognized that small-scale farmers in developing countries do make rational resource allocation decisions. The difference between them and resource-rich farmers is a question not of rationality but of objectives. From sheer necessity small-scale farmers operate highly complex farming systems with limited resources in an uncertain environment (both biological and institutional): this forces them to adopt a management system based not on 'profit-maximization' but on 'risk-aversion'. We, therefore, conclude that research programmes which do not specifically focus on small-scale farmers, and which are not based on an understanding of their complex farming systems, will not produce appropriate innovations for this group of farmers, e.g. the Green Revolution (Box 7) (Biggs, 1989; Boon and Bunders, 1987; Joffe, 1986; Lipton and Longhurst, 1989; Wolf, 1986).

A belief in the trickle-down effect as a mechanism for the dissemination of innovations is probably the main reason why small-scale farmers have not been considered as a specific target group for agricultural research. But there are others. Principal among them are the social background and career ambitions of scientists in developing countries (Chambers and Ghildyal, 1985). Almost by tradition, many agricultural scientists find it difficult to appreciate the circumstances of small-scale farmers. Scientists often come from relatively rich families: they are usually based in towns or cities and their experience of rural areas is mainly limited to those near the cities. There they are most likely to encounter large-scale farmers who grow produce for the urban populations. Their contacts, therefore, are with farmers of higher status, more influence, greater wealth and better education. Furthermore, from the standpoint of professional advancement

Box 7:
The Green Revolution

The cornerstone of the Green Revolution was the development of new, improved seeds. Scientists in the late 1960s perceived that the newly developed high yielding varieties (HYVs) of crops could produce a 'revolution' in Third World food production. The results of the Green Revolution have indeed been impressive in many areas, especially in Asia. In many regions, HYVs doubled or tripled food production (per hectare per season) in 20-30 years, outpacing population growth. History records no increase in food production that even remotely compares in scale, speed, spread and duration with that of the Green Revolution.

However, the same Green Revolution had dramatic negative implications for many small-scale farmers. The yields of HYVs are considerably higher than those of most local varieties. To achieve high yields, however, careful management methods and relatively high and regular applications of fertilizer, pesticides and water were required. Farmers without access to these inputs did not and do not benefit from the new seeds. In most African countries, for instance, these inputs are both scarce and subject to year-to-year fluctuations. The farmers still use the seeds but, over the years, yields have dramatically decreased. For example, cotton production in southern Chad initially rose from 90,000 tonnes per year in the early 1960s, to 120,000 t/y in 1972 and 155,000 t/y around 1977. Many farmers could, however, not afford to apply sufficient fertilizer: between 1972 and 1975, only 30-50% of the crop's nutrient consumption could be supplied as fertilizer. Inevitably, this lead to nutrient depletion in the soil. Consequently, from 1976 onwards, cotton production dropped dramatically, reaching levels as low as 68,000 tonnes per year in 1981 and recovering only slowly thereafter.

The introduction of HYVs put pressure on the traditional farming system: traditional methods of preserving the sustainability of the farming system - the use of fallow periods, traditional (multi-) cropping patterns, traditional systems of maintaining soil fertility, the use of trees - have disappeared in many areas. Soil degradation and soil erosion are widespread. The HYVs replaced many traditional, locally-grown varieties of crops leading to a loss of valuable genetic resources. Sustained, intensive farming under financially-limited conditions is every bit as complicated as traditional methods.

Currently, about 40% of rural populations in developing countries cultivate HYVs as their major crops. Yet, except in East Asia (including China) the poor in these 'Green Revolution-regions' have become poorer, either in absolute terms or, at least, relatively. Africa benefited least from the Green Revolution. Improved maize varieties and hybrids have boosted harvests in Kenya, Zimbabwe and South Africa. But, on the whole, the Green Revolution did not bring a decisive change in the continent's food outlook.

and recognition, it may not be in their interest to focus research on small-scale farmers' problems and concerns.

When these various factors are considered, there is little reason for scientists to cast doubts on the 'top-down' technology transfer model. It is 'more than understandable that agricultural scientists ... do not doubt that the transfer of technology model is appropriate for their work. They have good reason to embrace it and little reason to question it: it is from the resource rich who adopt, much more than from the resource poor, who do not adopt, that they get most of their feedback on the value of their technology' (Chambers and Ghildyal, 1985).

Both from the outcome of the Green Revolution and from a consideration of the relevant social influences on the agricultural research process, we must conclude that top-down technology transfer does not serve small-scale farmers well. Unless small-scale farmers are a specific target group for technology development, it is highly unlikely that technological innovations will trickle down to the small-scale farm level. The corollary of that is that in order to focus research on technologies appropriate for small-scale farmers, technology development must be based on a thorough understanding of their farming systems.

5.3 Farming systems research

Farming systems research (FSR) was born of the premise that technologies should be 'appropriate' to the targeted group. For this, a 'bottom-up' approach was needed to replace (or supplement) the top-down technology transfer model. Despite the diversity of terminology, methodology and scope of application used by its different 'schools', FSR is characterized by the identification of small-scale farmers' problems and needs based on a more complete understanding of their farming systems (CIMMYT, 1984; Collinson, 1988; Farrington and Martin, 1987; Richards, 1986; Simmons, 1986).

FSR gained widespread acceptance among research strategists and donor agencies in the late 1970s and 1980s. Donor organizations including USAID (United States Agency for International Development), the Ford Foundation, CIDA (Canadian International Development Agency), DGIS (Netherlands Directorate General for International Cooperation), ODA (United Kingdom Overseas Development Administration), FAO (United Nations Food and Agriculture Organization), EC (European Community),

IFAD (United Nations International Fund for Agricultural Development), and the World Bank have all funded FSR projects of various kinds.

Despite this support, the results of FSR have been 'patchy' and have certainly fallen short of the great expectations of its devotees (Collinson, 1988; Farrington and Martin, 1987; Richards, 1986). One of the most fundamental criticism of FSR is that, like the top-down technology transfer approach, it is still, in essence, a 'top-down' approach. The interaction between researchers and farmers is consultative; scientists consult farmers about their problems, but it is the scientists who develop 'solutions'. If one draws a commercial analogy, FSR is actually a tool to help research station scientists more accurately analyse the 'markets' for their own preferred technologies. It does not empower the end-users of the technologies to influence the design of research projects. Neither does it respect the farmers' abilities to innovate. Moreover, FSR is often perceived by the national agricultural research systems as something imposed from 'outside'. Strengthening indigenous institutional capacity and human resources was not the primary interest of FSR; large FSR programmes were usually led by expatriates, while local scientists and institutional networks were hardly involved. Furthermore, FSR considers farming systems as isolated sub-systems and has therefore too narrow a focus: this implies that broader interactions (e.g. cultural, political and economic) are not sufficiently taken into account. For these reasons FSR has not been able to influence the agenda of public sector research.

Thus it can be concluded that even the most thorough understanding of a farming system is not enough to realize the development of appropriate technologies. In order to guarantee that priorities in agricultural research are established from the bottom up, a form of partnership between researchers and small-scale farmers is necessary. Furthermore, farming systems should be considered within a broader cultural, political and economic context. In this way, one can hope both to influence biotechnology R&D agendas and anticipate external elements which might influence the execution of projects.

5.4 Participatory technology development

Participatory technology development is 'the practical process of bringing together the knowledge and research capacity of the local farming communities with that of the commercial and scientific institutions in an interactive way' (ILEIA, 1989). A number of leading proponents of participatory technology development have, for some time, pointed to the 'indigenous knowledge systems' of traditional societies (Chambers, 1990; Rhoades, 1984; Richards, 1986). Farmers have a wealth of knowledge of their environment, they have developed specific skills to use this environment, and are very active and creative in adapting it for achieving their objectives. Strengthening and supporting 'informal' R&D capabilities of farmers is crucial for genuinely sustainable and equitable rural technology development (ILEIA, 1988, 1989). The key concept is 'partnership', with the different partners working together to improve the available technology. Participatory technology development seeks to strengthen the existing experimental capacity of farmers and to enable sustained local management. It has a bottom-up orientation in the research process; its agenda is set from client priorities. Participatory technology development has many local and personal variants and has been given several names, such as farmer-back-to-farmer, farmer-first-and-last, farmer participatory research, people-centred agricultural improvement, and on-farm-client-oriented research.

A relatively rapid method for identifying end-user needs often used in participatory technology development is 'participatory rural appraisal' (Chambers, 1990). A multidisciplinary team makes short visits to the target area and gathers the data required. Participatory rural appraisal is based on an exchange of information and a belief that researchers are as much 'students' as 'experts'. Practitioners emphasize the need to avoid lecturing, or taking preconceptions into meetings with local communities. Researchers should be able to acknowledge their own ignorance and not worry about being 'wrong'. Participatory rural appraisal enables researchers and members of a community to describe the local farming system in a mutually comprehensible way. In sharing this information, the researchers are in a position to discuss and agree with local people on the most appropriate avenues for research, either at farm level or within local or international research institutions.

Participatory technology development is an integrated approach which takes into account the effects of agricultural practices on production and on the environment while focusing on locally available resources. However, it has focused mainly on adaptive research at local research stations and on on-farm research, largely ignoring the development of 'new' technologies like biotechnology. As a result, it has had little influence on R&D agendas.

Participatory technology development does provide some important pointers for the way ahead. It shows that it is possible to identify and design projects *with* small-scale farmers. It has demonstrated that farming communities are stores of knowledge which are highly relevant for the identification of appropriate agricultural research. Scientists are not the sole guardians of expertise: they can and should learn from farmers. Finally, participatory technology development has shown the merit of using multidisciplinary teams for the identification of end-user needs through cooperation with local farmers.

5.5   Conclusions

The 'top-down transfer of technology' model is inappropriate for the development of agricultural innovations for small-scale farmers in developing countries. 'Farming systems research' represents a first step in developing agricultural innovations based on the needs of small-scale farmers. Its successor, 'participatory technology development', achieved the inclusion of end-users in project design and implementation. In an interactive process, the knowledge of local farming communities is merged with that of agricultural research institutions. Even participatory technology development, however, is still deficient: it does not provide a strategy for systematically influencing R&D agendas.

Chapter 6

Implementation strategies for innovative projects

6.1   Introduction

If small-scale farmers in developing countries are to derive any benefit from biotechnology, it will be essential to systematically influence R&D and policy agendas. In Chapter 5, we concluded that neither top-down transfer of technology nor farming systems research nor even participatory rural development provided such a strategy.

In this chapter, we will draw lessons from analyses of innovative processes. We will start by examining entrepreneurialism in the corporate environment and will focus specifically on those innovative processes that require new structures and/or policies. Corporate change processes are often realized by a team, and this we believe, although with some modifications, to be appropriate for our purposes too. One of the changes is that in the case of small-scale farmers in developing countries, an interdisciplinary composition of such a team is a necessity. We will, therefore, examine how interdisciplinary teams can be built and how they function. In this respect, much can be learned from research on the management of collaborative projects involving both scientists and non-scientists - between universities and industries, and 'alternative cooperations'.

6.2 Entrepreneurship

Both the development and implementation of biotechnology for small-scale farmers are innovative projects which cannot follow the patterns of top-down planned, or incremental processes. Any innovation which aims at the improvement of the situation of small-scale farmers is, almost by definition, a threat to vested interests, and requires the building of new structures and/or policies. This situation is not uncommon in innovative processes in corporations and other organizations. To identify what is important in such processes, we need to understand how innovations actually take place. We will address both the specific activities and the context that are needed to realize an innovation. These two aspects were dealt with extensively in the literature on processes of change in corporations and other organizations in the 1980s (Box 8).

Box 8: Processes of change in corporations and other organizations

Authors such as Rosabeth Moss Kanter, Ronnie Lessem, Gifford Pinchot and Peter Drucker have examined the way in which innovations are implemented within corporate and other organizations. They describe how corporate entrepreneurs stimulate their organizations to experiment on uncharted territories and to move beyond what is known in order to improve their market position. Kanter focuses specifically on American business organizations (Kanter, 1983). Lessem shows that Kanter's analyses also apply to business organizations in Europe and Africa (Lessem, 1986). Drucker focuses on business organizations as well as on public institutions such as government agencies, labour unions, churches, universities, schools, hospitals, community and charitable organizations, and trade associations. In his book *Innovation and Entrepreneurship*, he concludes that entrepreneurship is by no means limited to the economic sphere, and that there is little difference between entrepreneurship in different organizations (Drucker, 1985). To a large extent, entrepreneurs in education and health care do the same things, use the same tools, and encounter the same problems as the entrepreneur in a business organization. His observations suggest that the same considerations apply to the realization of innovations for the poor in developing countries.

All these authors describe the sort of conditions which hamper innovation. For example, Pinchot states that the stagnation of innovation processes in large corporations is an almost inevitable result of organizational rigidity (Pinchot, 1985). He discusses this 'implementation crisis' and concludes that large organizations produce many interesting ideas but generally are unable to implement them. His recipe for solving this problem emphasizes the importance of: a vision; networking and team-building; synergy in the innovation project; the identification of sponsors; and the necessity to take into account the corporate 'immune system'.

Kanter focuses on 'segmentalism' as the barrier to innovation (Kanter, 1983). Segmentalism reduces integration of information available within an organization and thereby reduces the level of the kind of pro-active behaviour needed to initiate innovative projects. Another constraining factor may be the lack of one of the basic sources of organizational power that entrepreneurs need: (i) information (data, technical knowledge, political intelligence, expertise); (ii) resources (funds, materials, space, staff, time); and (iii) support (endorsement, backing, approval, legitimacy).

The various description of the characteristics of corporate entrepreneurs, of how they work and of which skills they have, show many similarities. Kanter summarizes the characteristics of entrepreneurs as follows: 'What corporate entrepreneurs show is a longer time-horizon, conviction of an idea, no need for immediate results or measures, and a willingness to convey a vision of something that might come out a little different when finished. And moreover, they have fun. Entrepreneurs are, above all, visionaries. They are willing to continue single-minded pursuits of a clearly articulated vision' (Kanter, 1983).

Knowledge-based innovations, like biotechnological innovations, involve, according to Drucker, high risks (Drucker, 1985). This puts a higher premium on foresight, both financial and managerial, and on being market-focused and market-driven when compared to other types of innovation. Knowledge-based innovation requires a careful analysis of all necessary factors; technological as well as social, economic and perceptual. Such an analysis can indicate whether these factors are available. On that basis, an entrepreneur can assess whether the 'missing links' can be forged, or whether the innovation should be postponed as currently infeasible. The absence of such an analysis is 'an almost sure-fire prescription for disaster'; either the knowledge-based innovation is not realized, or the innovator succeeds only in creating an opportunity for somebody else (Drucker, 1985). Such an analysis may be considered obvious. Yet, they are rarely performed by scientific or technical innovators, because they think they already know what is needed. They tend to be contemptuous of anything that is not 'advanced knowledge' and of anyone who is not a specialist in their own area. They tend also to be infatuated with their own technology, often believing that 'quality' means what is technically sophisticated rather than what gives value to the user. In this respect, Drucker says, scientists and technologists 'are still, by and large, nineteenth-century inventors rather than twentieth-century entrepreneurs' (Drucker, 1985).

Corporate entrepreneurs achieve innovation by working in a collaborative and participative fashion. They persuade rather than order, build a team (including formal task forces or committees), organize staff meetings, obtain and share information, use brain-storming sessions, seek input from others (including end-users), adopt the suggestions of subordinates and peers, show sensitivity to the interests of others, organize support and resources, and are willing to share rewards and recognition. The participative process is needed to gain support for the project and reduce its risks, guarding against its failure and encouraging its completion. The involvement of others is as a kind of check-and-balance system, reshaping the project so that it approaches its final goal - successful implementation. For the corporate entrepreneur, tolerance for uncertainty is more important than tolerance for risk.

According to Kanter a prototypical innovation, led by a corporate entrepreneur, has three identifiable steps, occurring sequentially or iteratively: (i) problem definition; (ii) coalition building; and (iii) mobilization (Kanter, 1983).

Problem definition

Problem identification is the first step in project definition. There may be many conflicts within an organization on the best method of attaining goals. Discovering the basis for these conflicting perspectives, and the gathering of hard technical data, at an early stage is critical. Active listening to the information circulating is the first step in generating an innovative project, and information is the first power tool. In the beginning, it is best to have as many sources as possible. As part of the process of translating a set of vaguely expressed needs into a realistic proposal for innovation, an entrepreneur will benefit from welcoming and canvassing a range of opinions from a broad cross-section of sources. Entrepreneurs may themselves examine the issue from a variety of perspectives, or they may seek the views of others. While gathering information, entrepreneurs can also 'plant seeds'; 'leaving the kernel of an idea behind and letting it germinate and blossom so that it begins to float around the system from many sources other than the innovator' (Kanter, 1983).

The early information acquired by the entrepreneur establishes the basis for the entire project. In this early phase, the technical and political information will usually have to be obtained from outside the manager's own work-unit. Even for entirely new initiatives, the technical information needed often already exists but must be collated from many fragmentary sources. Piecing together the political picture will help identify those groups with stakes in the issue and clarify other developments that might be tied to the project to help sell it and support it. Without political and technical information early in the project, it will probably never get beyond the proposal stage. There are two other vital kinds of information: that which demonstrates the need for an innovation, and that which makes the case that the chosen innovation (of all those possible) addresses the need best.

In formal surveys of clients or informal conversations, the entrepreneurial manager will begin to lobby and line up supporters for his or her own view. Information then serves the dual purpose of shaping project activities and of creating an instrument to persuade resource holders to back the project. From this point on, all the fragments of information begin to be focused on and directed towards a particular target. Entrepreneurial managers need to have thick skins - to develop personal immunity to the comments that the kind of thing they have in mind 'cannot be done, has never been done, or was tried and did not work'. By now, the manager will have a specific project in mind, one that embodies a vision of possible results and that encompasses the interests of the other stakeholders.

Coalition building

When the project has taken shape, it must be sold to would-be backers. The initial, rather vague, assignment is usually not automatically backed by financial and other resources required to implement the project. Innovators, therefore, have to be team creators as well as team users; they may work through a formal task force, or they may build an - informal - network of allies. Pinchot gives some ideas on how such teams can be formed. 'The best teams seem to grow by accident rather than by active planning. The practice of "bootlegging" a new project with volunteers before it is approved provides an ideal environment for growing a team. During the bootlegging phase you ask for help. Some people are drawn to your vision and your leadership style and get things done. Others don't work out. When the time comes to formalize the project, you will have sorted through a fair number of potential team members at a very low cost, both financially and emotionally' (Pinchot, 1985).

What is critical for the success of the innovation at this point is coalition building. A 'coalition' of people from different areas serves as a check-and-balance. This is important because the activities in innovative projects are, by definition, non-routine and not subject to the usual controls or protocols. Coalition building also ensures that sufficient and appropriate support will be available to maintain the momentum of a project and guarantee its implementation.

An important part of coalition building is establishing 'cheer-leaders' -

supporters from higher levels - who can be 'lined up' in advance of formal approval. Thereto, peers, managers, stakeholders, potential collaborators and, sometimes, even customers will be approached in one-on-one meetings. Those approached have a chance to influence the project while the innovator has an opportunity to 'sell' it.

The next, 'horse trading', phase is essentially the bartering of expected pay-offs from the project in exchange for support, time or money. In the coalition-building phase, the supporters are not merely passively involved; their comments, criticisms and objectives help to redefine and shape the project into one that is more likely to succeed. The result of the coalition-building phase is a better information base and a set of 'reality checks' which ensures that projects unlikely to succeed will go no further.

Mobilization

With a team ready to carry out the specific steps needed for implementation, the entrepreneur's job is, primarily, to act as a buffer between the project and the rest of the world. The project leader's part involves maintaining the boundaries and integrity of the project by maintaining momentum and continuity of team effort, adapting the organizational structure needed to ensure implementation, and seeing that external communication paves the way for a receptive environment. Strikingly little opposition is encountered in this phase, probably because early success at coalition building determines whether a project starts at all.

Conclusion

Drucker made clear that the implementation of innovative ideas in industry and in other organizations are comparable (Drucker, 1985). The ways problems are solved seem to follow general patterns. Successful innovation processes consist of three steps:

1 . Problem definition: an entrepreneur with a vision and with an idea, who is capable of evading corporate resistance to change, makes use of opportunities the context provides (synergetic effects), and acquires and applies information (statistical data, technical knowledge, political intelligence, expertise) to formulate a realistic innovative project.

2 . Coalition building: the development of a network of backers who agree to provide resources (funds, materials, space, staff, time) and/or support (endorsement, backing, approval, legitimacy).

3 . Mobilization: the investment of acquired information, resources and support in the project itself. In this phase, the projects' working team is activated to bring the innovation from idea to practice.

Essential elements in the implementation of innovative ideas are 'the teams'; the task force - the formal team - and the network - the informal team. It is the formal team which is responsible for the elaboration of an idea; it is within the teams that information is generated and exchanged; and eventually, it is the teams which formulate solutions in the light of the constraints and opportunities they discover.

6.3   Interdisciplinary   research

Developing biotechnology for small-scale farmers by influencing R&D agendas requires teamwork at two levels at least; informal and formal. The informal team is, in essence, the network of information and resource nodes which will be necessary throughout the entire project. The formal team is responsible initially for planning and ultimately for implementing the project. It is the formal team which initiates and mobilizes the larger informal one.

By necessity of the project envisaged, the formal team should be interdisciplinary, including both technical and social scientists, and the individual members should be able to collaborate productively. Implementation of interdisciplinary research is a recognized problem; many pitfalls often lead to disappointing results. In the literature, a remarkable consensus both on the problems of interdisciplinary research and their solutions exists. The following, often related, problems, seem to occur:

1 . Discipline-centrism (regarding the own discipline as the 'right' one).
2 . Perceived inequality in status between disciplines.
3 . Communication problems caused by jargon.
4 . Conflicts of personality.
5 . The output is not considered to be of academic interest.

Moreover, these problems, although common, are generally ignored (Klein, 1986). The common solution put forward is to recruit 'bridge scientists' or 'intermediaries'. The role of a 'bridge scientist' is often to initiate the project, to translate research problems in terms comprehensible to members of each discipline, to enhance the interdisciplinarity without imposing any authoritarian view, and to take charge of (or, at least, be strongly involved in) the integration of information. Of these roles, the most critical concerns the promotion of interaction between people; the bridge scientist is more a mediator between people than a translator of concepts (Anbar, 1986). According to De Zago, collaborative projects benefit from team members sharing the following traits: openness, respect, persistence, trust, and willingness to express oneself (De Zago, 1986). Conversely, rigid attitudes towards research design, inability to communicate, timidity and irascibility reduce the chances of successful teamwork. An interdisciplinary team consisting of bridge scientists would have a high chance of realizing results. This in itself, however, is insufficient to realize an interdisciplinary project. Group members should also share a clear mission to which they are strongly committed. Wilbanks suggests that members of interdisciplinary teams should: have a mutual focus, review each other's draft materials, take joint responsibility for written reports and be in physical proximity (sharing offices, performing field trips) (Wilbanks, 1986).

We conclude that bridge scientists are likely to play as crucial a role in developing biotechnological projects for small-scale farmers in developing countries as in other interdisciplinary/collaborative projects. Their shared commitment to a clearly directed mission is important.

6.4 Collaborative projects between scientists and non-scientists

If interdisciplinarity in research is a source of difficulty, what of a wider interdisciplinarity: the cooperation between scientists and non-scientists. That is certainly pertinent in any discussion on biotechnological innovations for small-scale farmers in developing countries. In this section, we will examine university-industry cooperations. There seems to be a natural partnership between the two sides to which much of the success of collaborative ventures has been attributed. More pertinent to small-scale farmers in developing countries are alternative cooperations - cooperations between scientists and groups with little involvement in decision-making on R&D. We have compared the university-industry and alternative cooperations to try to understand how and why they succeed, or fail.

University-industry cooperations

Biotechnological research agendas of universities increasingly reflect the priorities of industry. How does industry exert its influence on policy and on academic research agendas? University-industry projects outnumber all other types of cooperations between university biotechnologists and external groups. It might be concluded (and often is) that industrial goals fit better with the goals of university researchers than the goals of other external groups. Industry-university cooperations are considered to be 'natural partnerships'. This naturalness is often attributed to three factors. Firstly, the financial resources of industry match academics' (increasing) need for financial support. Secondly, there are no cognitive barriers: scientists whether in industry or universities use the same jargon. Thirdly, there is a structural fit: industry's demand for technical solutions is compatible with, and not necessarily competitive with, the scientific goals and approaches of university scientists.

However, these explanations do not withstand rigorous examination. Some cooperations can be explained in this way but our analysis indicates these to be exceptions to the rule. On the question of funding, for instance, industry is not always the major provider of funds in university-industry cooperations. Many cooperations are mainly financed by the government. Such support is justified on the basis of maintaining national competitivity in the (bio-) technology race (Roobeek, 1990). Universities and industries alike will pay particular attention to factors which might influence the climate for subsidizing specific research topics (e.g. new social problems or issues on the political agenda, change of government, international trends, etc.). We would conclude that it is often not the availability of industrial funds which leads to industry-academic collaborations, but the capacity of industry for organizing the necessary financial resources.

The absence of cognitive barriers as an explanation for the prevalence of university-industry cooperations is also questionable. In discussions of pure science, the jargon may be shared, but concepts are not. In any case, shared jargon is only relevant during the realization phase of collaborative projects. This is only the most visible part of a complex process. Those involved in the, important though often invisible, preparatory phase (representatives of universities, industry, ministries and often of applied research institutes) are usually specialists in narrow fields but are laymen in others. Consequently, the preparatory phase focuses on the generalities of the outcomes of research (e.g. on the development of a certain technique or modification of a plant or micro-organism) rather than on the details of the research techniques. Thus in the decision-making process that precedes research funding, the cognitive barriers are less important than the exchange of information, and the development and assessment of ideas. Of course, the exchange of information is facilitated if there are no cognitive barriers.

Finally, is it really true that industry and universities have comparable goals? On closer examination, the nature of industry's demands does not appear to mesh with the goals and approaches of university scientists. In many cases, cooperation starts with a vague and rough idea (an opportunity seen by, for instance, a scientist, an industrial manager, or a politician). This vagueness, however, need not be an obstacle to successful cooperation. In the preparatory phase, ideas which originally fall outside current frameworks for support are transformed so that positive decisions are more likely. Projects and programmes are not simply identified and selected - they are actively constructed. These activities take place in an informal atmosphere. The underlying motivation of each player to participate in such an interactive process may be different. What is crucial is that each is committed to the same central idea, and that together they have a good overview on the constraints and prevailing conditions under which the idea will flourish (Blume, 1987). This interactive process between the different players leads to the formation of an informal team which supports and further develops the idea into broadly justified projects. Such an informal team is crucial for integrating information from various sources.

Furthermore, it is differences rather than similarities which often trigger cooperation. Cooperation with external groups provides university scientists the opportunity of conducting research which is not considered 'interesting' when judged against the traditional norms and rules of the scientific community, but which is nevertheless crucial for the field concerned (Blume, 1987; Bunders, 1987). From industry's point of view, cooperation is often an escape from rigid structures of the organization. Narrow strategic focus, bureaucracy and high uncertainty about the feasibility of a project often prevent companies from developing an idea completely within their own organization (Schmitt, 1985). Risk-sharing with organizations like universities can make a project more presentable within a company.

In sum, when the 'naturalness' of university-industry partnerships is examined more closely it is not quite as natural as one might expect. Our analyses show that industrial resources, absence of cognitive barriers and a structural fit are not conditions for effective cooperation. Many university-industry relationships succeed because a rational, well-considered approach is taken in identifying, specifying and realizing opportunities, organizing financial support, and bypassing rigid scientific and social structures. University-industry research programmes are the result of interactive processes based on mutual interest and shared commitment.

As with entrepreneurial team-building, three main activities appear to be crucial in the preparatory phase of a collaborative project. The first is the discussion of an idea with potential participants to gather information, to specify the idea, to solicit commitment to the idea and to exchange information. The second is to establish an informal team consisting of committed representatives of the organizations. The third activity is to specify and justify the idea on the basis of the broad information which is available in the informal team, thereby taking into account the actual potential for change.

Alternative cooperations

For several years, we have studied collaborative projects between scientists and grassroots organizations and gathered information from numerous sources on their experiences. Examples are cooperations between scientists and environmental groups, small-scale fishermen, labour unions and interest groups in developing countries. One of the major problems encountered in alternative cooperations is that the proposed projects and programmes are seldom implemented. Often, research proposals are insufficiently specified and poorly justified, making it easy for decision-makers and/or scientists either simply to ignore project proposals or to ridicule them as naive and unrealistic.

There are several reasons why project proposals fail. The quality of the input is one: sometimes non-scientists may be unable to provide the information necessary to specify a project (Leydesdorff and Van den Besselaar, 1987). Inter-organizational in-fighting is another obstacle: conflicts between and within different grassroot organizations hamper the justification and specification of projects (Cramer *et al.*, 1987). Furthermore, grassroot organizations are often reactive: they criticize existing research projects or programmes which they find unacceptable, instead of focusing on the development of alternatives. Another constraint on effective cooperation is the interdisciplinary character of many research proposals.

Some alternative cooperations successfully pass through the preparation phase, but encounter problems during implementation. Those problems are often associated with the diversion of a project from its original course. As a result of 'hidden agendas' or the 'honeymoon syndrome' (Wagner, 1987), projects and programmes which have been initiated for a specific socially relevant purpose or to empower deprivileged groups are later hijacked or compromised by other, often more powerful, groups. This phenomenon seems to be more the rule than the exception.

Several structural solutions to such problems have been suggested and sometimes realized (Bunders and Leydesdorff, 1987). One, which has occurred in the Netherlands, is to establish 'science shops' within universities to ensure the access of groups without money and information to scientific developments. Science shops try to stimulate scientists to work for non-influential groups and to prevent research undertaken for these groups being ignored or ridiculed. Another structural solution is to institutionalize new scientific fields or to establish new institutes, which focus specifically on important social questions ignored by the traditional disciplines/institutes.

Although these solutions have proved useful for realizing alternative cooperations, to a large extent, they are not integrated into mainstream research activities. Science shops, new fields and new institutes may answer specific needs but the jaundiced mind might perceive that they were also a way of avoiding making any alterations to the basic structure of mainstream research. It is certain that they are insufficient to effect structural changes on public research agendas.

In the analysis of alternative cooperations conducted by the Department of Biology and Society, it was apparent that successful collaborative projects included an extensive preparatory phase. During this phase, possible (combinations of) barriers were carefully assessed and removed and/or avoided, after which a proposal was carefully prepared. Part of a successful strategy appeared to be the identification of a prestigious 'sponsor' who supports the project morally and catalyses the process. Another part is the use of publicity. In alternative cooperations, favourable public opinion is often an important weapon in combating the resistance of decision-makers. This use of the media may, however, make the project more vulnerable; this is particularily so when media support is sought before the project is sufficiently specified or when the project competes with more traditional approaches (which is almost always the case).

Another important strategy during the preparatory phase is networking (Box 9). The challenge is to specify the project sufficiently to make it both scientifically interesting and socially relevant. An intensive exchange of information through an informal network is often used both to model the research proposal and to attract various players to the project. Usually, there is no seed money for this activity: shared values and commitment to the idea compensate for funds.

Box 9: Networks

A network is a set of elements interconnected in a multiple way. Each element is connected to every other element and, through each other, to a centre. The network of interlinkages affects the flow and distribution of resources and power. Networks can be defined in terms of (i) the nature of their elements (e.g. individuals, departments, organizations, etc.); (ii) the channels connecting them (e.g. formal and informal lines of authority, information, decision, etc.); and (iii) the nature of transactions that can occur through these channels (e.g. consumption, distribution, or money-lending). Networks may vary in scope, complexity, intricacy, stability and flexibility (Suarez, 1984).

There are both formal and informal networks. Formal networks are relatively easy to identify since they follow formal communication channels. Informal or 'undercover' networks are less easy to identify, but are very important since many transactions and negotiations are conducted through such channels. Schon states: 'Informal networks have long served to enable people to get things done when the formal networks failed. The very life of social systems has depended on the operation of informal networks' (Schon, 1971).

There are several strategies to manage network relationships. This requires the development of ad hoc interlinkages, a process which involves design, creation and negotiation. The person who manages network interlinkages must have certain abilities to facilitate understanding and to persuade (Schon, 1971).

Successful alternative cooperations (Box 10; Von Gizycki, 1987) between scientists and non-scientists are often initiated and developed by scientists who share norms and values with the non-scientific groups. Such scientists are dubbed 'intermediaries' or 'boundary spanners' (compare with bridge scientists in Section 6.3). An important task of the intermediary is to channel information between the various players and to mobilize them; in other words, to establish an informal network. More importantly, perhaps, the discussions provide an opportunity to air mutual prejudices and preconceptions in an open and relatively non-threatening atmosphere. The aim is to progress towards mutual trust and empathy.

Box 10: Effective cooperation between scientists and non-scientists: the German Retinitis Pigmentosa Society

Retinitis Pigmentosa (RP) is an eye disease, still incurable, which eventually causes total blindness. In past decades, the German Retinitis Pigmentosa Society (a self-help organization of RP patients) formulated its own research interests and succeeded in influencing research agendas. The way in which it did this produces some useful points in our discussion of cooperation between non-scientists and scientists.

Soon after the foundation of the self-help group, it became clear that no systematic research was being carried out to find the cause of the disease. The ophthalmological community was relatively unstructured and not specifically interested in RP research, also because of earlier failures. This provided an 'entrance' to the scientific community for new demands: the scientific community was receptive to new ideas.

A key person in influencing and promoting research appeared to be the 'research-referee' appointed by the self-help group. Being a patient as well as a scientist implied that he was in a good position to play a dual role. Acting as a channel of information from the scientific community to 'laymen', he enabled the self-help group to take heed of scientific developments as it formulated its own research priorities based on patients' needs and requirements. On the other hand, as a patient familiar with 'scientific thinking', he could convince the scientific community of the importance of the priorities formulated by patients. The research-referee enabled the self-help group to formulate a framework under which research might take place more effectively. This led to a number of activities; the publication of a report on the international status quo in RP research; the formulation of patient demands for the research community; the organization of international conferences; the development of an integrated RP research programme; and the granting of a RP research prize.

It was the self-help group which organized the first RP congress as a way of bringing its research priorities to the attention of the scientific community. At this conference, a large group of researchers in the field of RP met for the first time. The personal contacts led to a feeling of common interest in RP research, which later became the motivation for establishing a scientific advisory board. After intensive preparatory work, this board defined the research programme requested by the German Retinitis Pigmentosa Society as a basis for funding future RP research work. The presence of an official from the Federal Research Ministry at the RP congress lent support to the decision of the Board to develop such an interdisciplinary programme. Before a scientific audience he stressed the Ministry's willingness to support such a research concept.

It became clear that the priorities of patients and scientists were not mutually exclusive. On the contrary, the patients' demand for an interdisciplinary approach to find the cause of RP appeared to be a cognitive challenge and a social opportunity for researchers. Interaction between researchers from different disciplines and between researchers and patients underlined mutual interests.

The preparatory process usually involves more than two groups. Information exchange between many relevant players makes it possible (i) to anticipate future problems in implementation; (ii) to identify synergetic effects, for example by making use of changes in the environment; and (iii) to understand how a programme can be embedded in existing policy. Broad criticism enhances the robustness of a programme: when an aspect of it is opposed, that opposition will be countered by evidence, alignments, interests and values. The integration of these comments into the proposal lends robustness. This approach also reduces the chance that a programme is ignored or hijacked.

Comparison of alternative cooperations with university-industry cooperations

Effective alternative cooperative research programmes, just as university-industry cooperations, are almost always the result of an interactive process in which shared commitment and mutual interests are essential components. In both types of cooperation, rigid structures constrain innovation and reduce the chance of success because, by definition, there is no routine way of dealing with problems. In the preparatory phase of both, opportunities are forged and a rough idea is specified by an informal team of players. Subsequently, to prevent neglect, ridicule and the emergence of hidden agendas, a project proposal must be made more robust through careful specification and legitimization.

One difference in the strategies for implementing projects is the importance of publicity. In alternative cooperations, favourable public opinion is often an important weapon in combating the resistance of decision-makers. Furthermore, in alternative cooperations, the professional assistance necessary to specify a project is usually not present and has to be hired. There usually are, however, no resources to pay for this expertise.

6.5   Conclusions

Analyses of various innovative projects demonstrate that the successful implementation of innovation depends on:

1 .   The vision of entrepreneurs and the opportunities they see.

2 .   A formal interdisciplinary team of intermediaries to initiate projects which involve scientists from a range of disciplines.

3 .   Networking and team building through informal channels.

4 .   A carefully designed preparatory phase in which a rough idea is further specified, legitimized and justified.

5 .   A prestigious sponsor to give moral support to the idea.

Chapter 7

The interactive bottom-up approach

7.1   Introduction

In Part One, we have shown that an understanding of farmers' problems, their interests and their farming systems is crucial in the assessment of the appropriateness of biotechnological innovations. This implies that access to information about this target group is indispensable. Another guideline is that innovations have to be feasible - that is, the expected output has to be achievable both scientifically and technically and in terms of its diffusion to farmers. The implication is that also access to information on scientific developments and (policies for) diffusion of the results is crucial. A combination of appropriateness and feasibility makes an innovation implementable. A third guideline is that a biotechnological innovation has comparative advantage in implementability, problem-solving capacity and cost effectiveness over other options.

Then the question becomes how to identify, formulate and prioritize appropriate and feasible biotechnological innovations for small-scale farmers which have a comparative advantage. Access to information and information exchange are crucial to answer this question with the aim of incorporating the priorities of small-scale agriculture in biotechnology research policies of governments, research institutions, donor organizations and other involved parties.

In current approaches priority is usually given to those projects which best fit into prevailing structures or to those environments holding the greatest promise for new development actions. There are, however, substantial drawbacks to these somewhat conservative approaches. In the first case, the status quo is assumed to be inviolate, and the few proposals that fit into these structures are unlikely to contribute substantially to the improvement of the position of small-scale farmers. In the second case, those areas that are already the most highly developed ones are usually chosen for new avenues for development. With this approach, it is unlikely that small-scale farmers will be selected as a target group. Rather it will lead to the promotion of large-scale commercial agriculture and industry.

From our discussion of entrepreneurship (Chapter 6) it is clear that innovation, in general, meets resistance because it threatens vested interests. It causes competition for scarce resources and has the potential to change (power) structures. This holds true for all types of innovation in all types of organizations. Thus it can be concluded, that aiming for a structural fit is, by its nature, an obstacle to innovation.

Different elements from the approaches discussed in the previous chapters are crucial for the incorporation of the demands of less powerful groups into scientific research. The first element is the importance of the preparative work. As we have shown in Chapter 6, an extensive and carefully designed preparatory phase involves the specification, legitimization and justification of a rough idea of an entrepreneur by the interaction of two teams (formal and informal). Such a preparatory phase will anticipate and deal effectively with structural problems. Another element of realizing innovation is implementation; bringing the idea into practice. The acquired information, resources and support are invested in the project itself and in restructuring the environment. Again this depends on the activities of a relatively small number of people.

As we have seen in alternative cooperations, it will be rare that in the case of biotechnological innovations for small-scale farmers, the activities defined in the preparation phase will be sufficient to implement them. Therefore, an additional element is required as a second phase. That phase is one of exposing the proposals to a wider, critical audience: a public debate is needed during which defined projects are discussed with both laymen and different types of experts. These discussions are intended to gain support, to reveal complications previously overlooked, and to assess their impact on the appropriateness and feasibility of the innovation. In this chapter, we discuss in more detail the three-phased 'interactive bottom-up approach'.

7.2 Phases and activities in the interactive bottom-up approach

The interactive bottom-up approach is divided into the following phases:

1. *Preparation:* In this phase, ideas are elaborated. In our case, the idea is that biotechnology can contribute to sustainable rural development. The output of this phase is the formulation and prioritization of research areas, projects and guidelines for the construction and assessment of new projects.

2. *Public debate:* The output of the preparatory phase is discussed widely and openly in order to gain support and to anticipate negative side-effects, constraints and synergies relevant in further developing the innovation. It may lead to the rejection of (part of) the ideas, or to a change of priorities and adaptation of the proposals. Ways of achieving a receptive environment for the effective implementation of the innovation will be discussed.

3. *Institutionalization:* Projects are more highly specified, programme studies are initiated, existing institutional frameworks are adjusted, and new institutional frameworks are developed.

The three phases are only partially chronological. Proposals may pass through the different phases more than once. Depending on the context of the country, the time for preparation and debate may be 5-10 months. Institutionalization may take several years. In the following sections, we will describe some of the specific activities that should be undertaken in each phase.

Preparation phase

Activities which take place in the preparation phase are:

1. Establishing a formal interdisciplinary team to catalyse and support decision-making on biotechnology for small-scale farmers.
2. Preparing an overview of relevant literature.
3. Generating information through interviews.
4. Establishing an informal team.
5. Exchanging information within the informal team.
6. Integrating results by specifying and justifying the ideas.

To give a clear view of what is required, we will discuss the relevance of these activities, how and by whom they are to be executed and the skills that each demands. We will also describe some of the problems which may arise and their possible solutions.

Establishing an interdisciplinary team to catalyse and support decision-making on biotechnology for small-scale farmers

The first step in making sure that opportunities for small-scale farmers in the field of biotechnology are not left unexplored, is the establishment of a formal, interdisciplinary team. Its main task is to ensure that activities are well planned, prepared and effected. The training and skills of the team members are crucial for the success of their activities.

Decision-making on biotechnology for small-scale farmers is a very complex issue involving different scientific fields. In an ideal situation, information from all relevant scientific disciplines is available within the team. At least three areas of expertise are required: biotechnology, technology assessment and development studies. As well as training and direct experience, the team members must have or acquire other, perhaps more important, skills. They must be skilled in intercultural interaction and rate highly in openness, flexibility, respect, the ability to listen and initiate, and in maintaining relationships (Hawes, 1980). A further skill of utmost importance is the ability to deal with (or have tolerance for) ambiguity. A consideration of successful alternative cooperations makes it clear that team members must have a strong commitment to and belief in the mission of the project. Another important characteristic of team members is the combination of a scientific background with a concern for the case of small-scale farmers so that they can act as intermediaries or 'boundary spanners' between the scientific society and the small-scale farmers (Mintzberg, 1979).

Preparing an overview of relevant literature

The exploration of ideas of biotechnological innovations for small-scale farmers in a particular developing country starts with a comprehensive literature overview. This should encompass information on the problems and interests of the small-scale farmers and on the links between their activities and other groups. The overview should address the national context and the agricultural sector in order to get a rough idea of the biotechnological solutions that are feasible and of the prevailing conditions that need to be taken into account. It should describe the relevant actors and institutions. The overview is compiled by the members of the interdisciplinary team. Literature assistants could help with gathering information.

The problems which may arise in the preparation of such a document are many. In the first place, information may be limited or difficult to access. Insufficiency of information may cast serious doubts on the whole exercise. Even before the first hurdle is crossed, one may wish to postpone the project and direct one's energies elsewhere. Theoretically, the availability of abundant information is an indicator that good assessments can be made. Conflicting and contradictory information may arise as different sources are consulted. Different authors will probably have different perceptions of the problems of small-scale farmers and their solutions. In the preparatory phase it is important to use as many sources as possible and to collate the different views rather than aiming for a unified conclusion. In this stage it is necessary not to describe all the information in detail but rather to indicate its nature and to summarize the most relevant general conclusions.

Generating information through interviews

Interviews are a way of filling gaps in the often fragmentary information from the literature review and of resolving some of the contradictions in it. Many different sources should be interviewed, particularly those which can provide information which is least likely to be found in published sources - the small-scale farmers and representatives of relevant institutions. There are many institutions which have information on small-scale farmers or which are or could be involved in decision-making on or implementation of innovations for small-scale farmers. The interviews are conducted by the members of the interdisciplinary team. The main topics to be discussed include:

1. Local initiatives related to rural development.
2. The main problems of the small-scale agricultural sector.
3. Existing or preferred options in solving these problems.
4. Ideas on possible contributions of biotechnology.
5. The possible side-effects and expected risks.

The interviews should be conducted in a way that avoids technology-push or creates false hope. Instead of starting the interview with the possible role of biotechnology, the interviewer should start with subjects like the main problems in small-scale agriculture, their origin, context and possible solutions. The appropriateness and feasibility of the possible solutions should be extensively explored. In discussing their feasability it is very important to give realistic information on the time-frame of the introduction of a new technology. Specifically when discussions are held with farmers themselves, care must be taken not to sketch a too optimistic picture of the 'promises' of biotechnology.

The interviews also serve a second purpose - as a forum for exchange of information. Decision-making on biotechnological projects and rural development will take place within disparate groups: farmers' organizations, women's organizations, governments of developing countries, donor organizations, non-governmental organizations and biotechnologists. All these groups possess specific relevant expertise in one area but lack other essential knowledge. Direct information exchange between groups is usually very limited. During the interviews, the interviewer can manage the exchange of ideas and interests of the different parties. The interviewer will act as messenger, a facilitator of informed decisions. This may help reduce prejudice about the intentions of other parties and increase the awareness that other people share the same commitment. Policy-makers and scientists can learn of the ideas and wishes of farmers' and women's organizations, and about their specific development activities. Equally, farmers' and women's organizations can learn about biotechnology and agricultural research activities, and their possible contributions to small-scale farmer's problems. This information can be used to formulate their own priorities.

The interdisciplinary team may encounter the problem that representatives of many groups are simply not interested in the problems of small-scale farmers. This is unsurprising since poor small-scale farmers usually receive little attention. An effective solution to this problem is to ask the interviewee to identify those people within the organization who are already committed to the case of small-scale farmers. In some developing countries, it may not be possible to identify such people and this will seriously hamper the process of identification, formulation and prioritization of biotechnological innovations.

Establishing an informal team

By now, members of the interdisciplinary team have collected much information which may still be incoherent. Processing of incoherent information is extremely difficult. There will be much information on why things do not work and why they never will without major structural changes. Usually there is no consensus on what solutions are feasible. For the members of the interdisciplinary team it is difficult to weigh the different solutions against each other. To deal with these problems, an informal team of people, committed to the improvement of small-scale farming should be established. The informal team consists of representatives of relevant institutions for agricultural development. Potential members will already have been identified during interviews. In this team, the information gathered so far by the formal team can be discussed, at first, in one-to-one meetings and, later, in meetings with increasing numbers of team members.

Exchanging information within the informal team

The members of the formal and the informal team have completely different backgrounds, skills, expertise and so forth. They share the idea that biotechnological research for small-scale farmers is possible and should be encouraged. It is the combination of the differences and the shared 'sense of mission' that makes collective and constructive discussions possible. The discussions will be structured on the basis of draft reports written by the interdisciplinary team.

Discussions should focus on opportunities and on how to deal with constraints rather than trying to define those constraints more precisely. Different group members will perceive constraints differently. Even the basic idea can be criticized. In the discussions concerted efforts are made to deal with the perceived problems. Earlier experiences with projects in which changes for the benefit of small-scale farmers have been implemented (within the university, the government, the farmer's organization, etc.) will provide important lessons. They may indicate under which circumstances which changes can be realized. This information is necessary to understand why certain 'solutions' must be discarded. In many respects, this exchange of information resembles the first phase of an interdisciplinary research project.

The first discussions may indicate that more specific information on certain issues is needed. Local consultants can be hired by the formal team to collect such information. They can often do such a job easier and faster than the team members.

Integrating results by specifying and justifying the ideas

The huge amount of information available from local consultants, literature overviews and interviews must result in an integrated view on the role biotechnology should play in rural development. At one extreme, it will be concluded that biotechnological projects simply cannot be justified. This would imply that virtually none of the conditions necessary for successful introduction of biotechnological innovations pertain or could be met within the foreseeable future. In other cases, however, some biotechnological innovations will be found to be appropriate, feasible and have a comparative advantage. The remaining challenge is to use the wealth of information gathered to establish and justify a prioritization of the problem areas to be dealt with by biotechnological innovations. Looking for synergy is very important; an innovation is more likely if synergy with other activities can be demonstrated.

Public debate

The result of the first phase is a specified and coherent view of the role of biotechnology in serving the small-scale farmers in an appropriate and feasible way. However, the results of these activities need to be reviewed and discussed not least by those interviewed. They must have the opportunity to criticize the analysis. After all, only those in the formal and informal teams will have had an overview of the full range of opportunities and constraints. It would be misleading to present the integrated results as a consensus document when those who have contributed do not get the opportunity to react. In any case, iteration of analysis may engender new contributions. The fact that many people have been involved in (parts of) the process does not ensure that all opportunities and problems have been identified. Furthermore, a wider discussion is a way of legitimizing the findings of the preparatory phase. Although the first phase may produce a robust proposal, a public discussion will enable broad consensus to be reached. Given that the small-scale farmers are usually not capable of attracting attention to their situation, a public debate will contribute to the visibility of their needs and problems and arouse public support. This will enhance implementation.

Organization of the public debate

During the debate, laymen and experts must have a chance to comment.

Opening a public debate with the presentation of laymen and non-influential groups, rather than those of the 'experts', makes clear that their contributions are valued. At the same time, the presence of highly prestigious people is also of utmost importance; it indicates that the issue is considered relevant. This may influence the nature of the debate positively. The public debate stimulates informal contacts through which information can be exchanged and new cooperations built. In this way, the informal team can readily be extended.

Questionnaires

To obtain information also from those who have not had a chance to express their opinions and ideas during the public debate, detailed questionnaires on the major topics discussed should be distributed.

Reporting

On the basis of all available information, a report is prepared which is presented to all those involved in the first and second phase (feedback). It is important to give the report a wide distribution, e.g. to send it to many organizations, so that follow-up will be more likely.

Phase of institutionalization

On the basis of the proposals prepared during the two previous phases, decisions can be made in different organizations. Farmers' and women's organizations, governments of developing and developed countries, international donor organizations, universities, multinationals and others may feel that that they would like to follow up on the ideas presented. They can prepare new projects, initiate programme studies, adjust existing institutional frameworks and/or establish new ones.

There may, however, be difficulties in coordinating the various efforts and the activities of 'entrepreneurs'. Each of them may have hidden agendas and only use the general support for the ideas to realize their own interests. An additional problem is that the the ideas of the report are not picked up. In the same way that 'trickle-down' to small-scale farmers does not occur naturally, the process of 'trickle-up' to implementing organizations needs specific stimulation. These problems necessitate further activities of the informal team. Team members will need to be vigilant in monitoring lack of coordination and abuse of the resources made available. They may need to take remedial actions or, at least, make others aware that deviations from the original proposals are taking place. In the complex contexts in which the innovations are being implemented, monitoring is essential to allow for adjustment without losing sight of the purpose of the original proposals. It will be important to ensure that the criteria of appropriateness, feasibility and comparative advantage of the innovations are maintained during implementation.

When the problems cannot be dealt with, it may be useful to organize a second debate to evaluate progress and to discuss constraints. Again, careful preparation of the debate will be necessary to ensure progression through the uncharted territory of small-scale farmer-led innovation processes. During the course of institutionalization, new insights may become apparent which may change the way projects and programmes are assessed.

Chapter 8

Concluding remarks

The problems of small-scale farmers are varied, complex and interrelated. Their solution requires an integrated approach which encompasses, *interalia*, gender issues, the rationale of traditional crop and livestock management, and the impact of the land tenure system on the stability and robustness of the farming system. However, as we demonstrated in Chapter 5, the implementation of research projects that take these aspects into account is still not very successful. Implementation of innovation is a general problem (Chapter 6) and one that is neither specific for innovations directed towards small-scale farmers, nor for other groups without key influence in policy arenas, nor for developing countries.

There are important lessons to be drawn from experiences with the implementation of innovations in various organizations. We have incorporated many of the lessons in the formulation of a model for a farmers'-led innovation process: the 'interactive bottom-up approach' (IBU approach). This aims to identify, formulate and prioritize biotechnological innovations for small-scale farmers in a developing country. A crucial aspect of the IBU approach is that information on the appropriateness, feasibility and comparative advantage of innovations is assimilated, discussed and integrated in projects and/or programmes.

An effective way to examine and influence structures and constraints surrounding the implementation of a biotechnological innovation is to establish an informal team of representatives of relevant institutions and groups. The team members may have little in common apart from their shared commitment to the idea that biotechnology can contribute to sustainable rural development. By means of their analyses, information and suggestions, it is possible to process the information gathered, and to reach an understanding of what is possible - what conditions can be fulfilled in order to implement proposed innovations. That discussion strongly influences the specification and legitimization of the results. A subsequent public debate can indicate whether these results are indeed appropriate and feasible.

With this approach, we have found that it is indeed possible, at a general level, to identify biotechnological innovations which are appropriate, feasible and have comparative advantage for application at the small-scale farm level in developing countries (for Pakistan, Broerse 1990-a; for Kenya, Brouwer *et al*, 1991; for Zimbabwe, Stolp and Langeveld, 1990; for Bolivia, Van Rijn, 1991). That does not mean that all biotechnological innovations are relevant everywhere. Their appropriateness, feasibility and comparative advantage are, by definition, strongly dependent on specific contexts.

It is pertinent to ask whether the IBU approach is useful as a method to deal with any local context. Assuming that it is not a universally applicable approach, can we identify where it might productively be used? In Chapter 7, we have already indicated some of the conditions that need to be fulfilled. Where, in institutions relevant for the innovation (i) it is impossible to identify people with a strong commitment to poverty alleviation in the rural areas, and/or (ii) where there is no information available on small-scale farming systems, it is not possible to use the IBU approach. The magnitude of the barriers between the elite and the poor varies strongly from country to country and is an indicator of the efforts that will be needed to establish an informal team. The often large geographical and cultural distance between those scientists and the small-scale farmers means that individuals committed to the improvement of small-scale farming are rare. It is important to note that it is not absolutely necessary that all people in the team are strongly committed. It is sufficient that a 'critical mass', willing to discuss and process the information and to contribute to the realization of an idea, is identified. The absence of information on farming systems is more difficult to overcome. Without this information, the IBU approach can only be used to formulate individual projects based on the guidelines discussed in Chapter 3 without endeavouring to influence R&D agendas structurally.

Perhaps the most important conclusion of this book is that scientific and technological developments can be oriented towards sustainable rural development in the Third World. Many of the elements for successful orientation exist; farming systems will have been described; there will be NGOs which are closely working with the small-scale farmers and who are willing to assist; there will be committed scientists in universities and governmental organizations; and there will be accommodating and helpful civil servants in governments who have been brought up in the rural areas. Sometimes, there will even be policies that can be used to legitimize activities and proposals. The major challenge is to integrate these elements. Taken as a whole, there are many reasons for moderate optimism in our ability to 'get things done', as the entrepreneurs might say.

Appendix

Agricultural biotechnology*

Jacqueline E.W. Broerse

Introduction

Throughout this book we have spoken much of biotechnology in the most general terms. Our justification is that the 'interactive bottom-up approach' (IBU approach) is not a technology-led approach; most of the work that the interdisciplinary team must undertake involves issues other than an appraisal of the technology itself. The range of biotechnologies that will be appropriate in a given small-scale farming situation will be considerably narrowed by the prevailing economic, political and infrastructural environment. When a proposal formulated by the IBU approach is completed, only a limited range of technologies will emerge as holding promise for improving the life of small-scale farmers. Then again, some donor organizations will only provide support for research in certain areas, both scientific and geographical, and this will inevitably further limit the technical avenues that can be explored.

It would, however, be counter to the entire ethos of the IBU approach to discount certain technologies at the outset. In this appendix, therefore, we will present a brief overview of many of the technologies that constitute agricultural biotechnology. The technologies we will be discussing are those which are widely dispersed among the research community. The purpose is to provide an overview of biotechnological techniques which

---

* Part of this appendix is taken from the book *Biotechnology for Small-scale Farmers in Developing Countries: analysis and assessment procedures*, Joske F.G. Bunders (ed.), VU University Press, Amsterdam, 1990, by permission of the publishers.

interdisciplinary teams embarking on the IBU approach in this field may use to initiate their search for technological solutions to identified problems. The overview is, of course, not a substitute for a deeper examination of the technical opportunities that should be conducted both through a thorough literature search and through direct contact with researchers.

Biotechnology is a continuum of technologies. It ranges from long-established and widely applied techniques to more strategic research on, e.g. recombinant DNA (rDNA) technologies of micro-organisms, plants and animals. The techniques are presented in order of increasing technical complexity: fermentation, microbial inoculation of plants, plant cell and tissue culture, enzyme technology, embryo transfer, protoplast fusion, hybridoma or monoclonal antibody technology and rDNA technologies.

Fermentation technology

For thousands of years mankind has been taking advantage of the activities of micro-organisms to produce foodstuffs and drinks without understanding the microbial (fermentation) processes. It was not until the latter half of the 19th century that attempts were made to standardize the fermentation processes in order to obtain products of uniform quality. The fermentation process can be divided into three phases (Greenshields and Rothman, 1986):

1 . *Preparation.* An organism with appropriate characteristics is acquired and selected. Subsequently, it is grown to provide an inoculum for the fermentor ('organisms growth container').

2 . *Fermentation.* Fermentation takes place in a fermentor which may be very simple consisting of a dug hole or small box, or a highly sophisticated large-scale, computerized production unit. It is convenient to distinguish two types of fermentations: (i) non-aseptic systems, where it is not absolutely essential to operate with strict monocultures and (ii) aseptic systems, where all micro-organisms but the one of interest must be excluded. Traditional brewing and waste water treatment are examples of the former while the production of antibiotics, amino acids and single cell protein are examples of the latter.

3 . *Harvesting and product recovery* (downstream processing). After growth and production phases, the micro-organisms have to be separated from the fermentation medium. This can be achieved by precipitation, filtration or centrifugation. If the required product is a component of the micro-organism, the organism must be disrupted and fractionated. Recovering components either from disrupted suspensions of micro-organisms or from the fermentation medium requires the application of techniques such as ultrafiltration, electrophoresis, chromatography, or affinity chromatography (a process whereby the specific binding properties of monoclonal antibodies or other ligands are exploited to remove just one type of molecule from complex mixtures).

The efficiency and yield of the processes can be increased at all three stages of fermentation through selection of more productive microbial strains, control of culture conditions, and improved purification and concentration. Depending on the configuration of the system, three main aims can be identified. In most traditional fermentations, for instance, the primary intention is to preserve the nutritive value of foodstuffs by preventing the growth of harmful micro-organisms; fermented products of this sort have long 'shelf lives'. The modifications of flavour and texture that accompany, say, yoghurt and beer making are in many ways incidental to the main purpose of producing an acidic or alcoholic environment, respectively, in which harmful organisms cannot thrive. Other fermentations, usually the more modern types, have been developed to produce specific microbial products or, nowadays, to use genetically engineered micro-organisms to produce molecules of non-microbial origins. The third category of fermentations is designed to degrade waste; there the aim is to use the micro-organism to convert organic materials to gases, inorganic material and water.

A recent important development has been the immobilization of micro-organisms. Microbes in immobilized form are more robust. They retain their metabolic activity for longer and have greater resistance to extreme conditions. Furthermore, when immobilized they do not have to be recovered for subsequent re-use (Greenshields and Rothman, 1986). Some current applications of fermentation processes are listed in Table 4 (OTA, 1984). Compared with chemical processes, fermentation has certain advantages but some disadvantages (Table 5) (OTA, 1984).

All fermentations require water because all micro-organisms require water. However, it is convenient to differentiate fermentations which take place in a liquid medium from those occurring on a solid phase where water is present in much lower amounts.

Table 4: Categories of current use of fermentation

---

1. Production of cellular matter (biomass), e.g. baker's yeast, single cell protein.
2. Production of cell components, e.g. enzymes, nucleic acids.
3. Production of metabolites, e.g. ethanol, lactic acid and antibiotics.
4. Catalysis of specific single-substrate conversions, e.g. glucose to fructose, penicillin to 6-amino-penicillanic acid, and stereospecific synthesis of fine chemicals (e.g. pharmaceuticals).
5. Catalysis of multiple-substrate conversions, e.g. biological waste treatment.

Table 5: Advantages and disadvantages of fermentation processes compared with chemical technology

---

Advantages:

1. The formation of complex molecular structures such as antibiotics and proteins where there is no practical alternative.
2. The selective production or degradation of one specific form of isomeric compound. Stereoisomeric compounds are chemically identical molecules which are mirror images of one another; usually only one isomer shows the desired biological activity - the other may be ineffective or even toxic.
3. Micro-organisms can execute many sequential reactions.
4. Micro-organisms can in some cases give higher yields.

Disadvantages:

1. Product yield in microbial processes can easily be influenced by contamination with foreign, unwanted microorganisms and changes in process conditions (pH, temperature, pressure, ionic strength, concentration of oxygen).
2. The desired product will often be present in a diluted complex product mixture and require separation and concentration which are highly energy-intensive processes. Although this is no problem in the case of high-value products like some antibiotics and rDNA products, for other products, the cost of purification and concentration may reach economically unacceptable levels especially when the product is of low value and present in a dilute form. For instance, the cost of bio-ethanol per tonne exceeds that of petrol by US$350 due to the high cost of dehydration (Senez, 1987).
3. Microbial processes are usually slow compared to conventional chemical processes.

Fermentation in a liquid medium

Fermentation in a liquid medium encompasses brewing, milk fermentations, industrial production of microbial metabolites and waste water treatment.

Alcoholic fermentation

The most important product of brewing is ethanol ('alcohol') - a metabolic end-product formed by, e.g. yeast when sugars are converted in the absence of oxygen (under anaerobic conditions). Carbon dioxide is also produced. Ethanol is commercially very important; not only is it the basis of the enormously varied alcoholic beverage industry (wine, beer, etc.), it is also a liquid fuel, an energy carrier.

Major problems in alcoholic fermentation in general are infection of the culture with unwanted bacteria and the production of acids. In controlled ethanol fermentations such as those used to produce fuel ethanol, however, the major constraint is that ethanol inhibits the growth of the micro-organisms that produce it; the maximum concentration in the reactors rarely exceeds 8-10%. Most of the rest of the fermentation medium is water which must be removed in order to be able to use ethanol as a liquid fuel. This is costly, contributing about 60% of the production price; dehydration of ethanol above 96% requires extremely energy-intensive chemical treatments (Greenshields and Rothman, 1986; Senez, 1987).

Milk fermentation

The main aim of milk fermentations, with products such as cheese, yoghurt and sour cream, is to preserve the valuable nutrients in milk. Cheese-making is initiated by the addition of a starter culture of lactic acid bacteria (usually species of the *Lactobacillus* or *Streptococcus* genus) to milk. The bacteria cleave the lactose to glucose and galactose, and these intermediate products are in turn metabolized to lactic acid. The acid coagulates the milk proteins, principally casein, to form a solid curd (principally fat and insoluble protein material) (Primrose, 1987). Acidification, removal of sugars and removal of water all serve to prevent the growth of most other organisms.

Milk is not a sterile substrate and the starter culture organisms have to compete with other micro-organisms. If they do not perform as expected, problems can occur. A major problem in milk fermentations is slow acid formation, which can result in off-flavours, extended process times or complete loss of production. This can be caused when the starter culture is inhibited either by bacteriophage infection or by antibiotic residues in the milk, or by the loss of plasmids (small pieces of DNA) which code for the ability to grow on lactose or to degrade casein (Primrose, 1987).

Production of amino acids

Another important area for fermentation is the production of amino acids. Almost all the amino acids used to complement plant proteins are obtained by fermentation using hyper-productive bacterial strains, particularly species of *Corynebacter*, *Bacillus* and *Brevibacterium*. Many plant proteins are only of limited nutritional value due to their lack of certain essential amino acids (like lysine, threonine, tryptophan) which mono-gastric animals (including pigs, young ruminants and poultry) are unable to synthesize. At present, the cost price of, e.g. lysine is still considered too high to compete with other sources of animal feed, although in Europe (both East and West) a number of large lysine production plants are being constructed.

Production of antibiotics and other secondary metabolites

Antibiotics are of higher value than amino acids. They are secondary metabolites (compounds whose synthesis occurs after microbial growth ceases) produced by micro-organisms, which inhibit or even kill other micro-organisms. To be effective, antibiotic molecules must interfere with a fundamental metabolic process of the micro-organisms without (ideally) affecting the human or animal host. The first antibiotic, penicillin, was discovered in 1929 and is produced by the obligate aerobic mould *Penicillium notatum*. The main antibiotic producers are fungi, actinomycetes (bacteria which grow in mycelial form) and eubacterials (Antebi and Fishlock, 1985). Much of today's antibiotic production involves semi-synthetic processes in which microbes are used to produce a precursor compound which subsequently undergoes chemical modification.

Appropriate micro-organisms are obtained by screening soil, plant or water samples for antibiotic activity; less than 1 out of 10,000 selected strains provides a useful active antibiotic (Antebi and Fishlock, 1985). When a suitable strain has been developed, it is cultivated in a fermentor, isolated and purified. If the biological quality and safety of the antibiotic is appropriate, strain improvement (through conventional mutation techniques or modern genetic modification) and fermentation development are undertaken (Antebi and Fishlock, 1985; Primrose, 1987).

Antibiotics are used for therapeutic treatment in human and animal health care. The major problem in the use of antibiotics is the frequent occurrence of resistance; the target organism acquires resistance against an antibiotic to which it was previously susceptible.

Apart from antibiotics, there are many other secondary metabolites with useful characteristics. For example, in human health care, anti-tumour drugs, cyclosporin (used for organ transplants) and substances which enhance the contraction of the uterus have been isolated from microbial sources. Similarly, in veterinary health care, microbial products include anti-coccidants, deworming drugs, growth promoters and muscle enhancing drugs, while in crop production (pest management), microbial compounds are used as fungicides, insecticides (e.g. toxin of *Bacillus thuringiensis*) and growth regulators (Antebi and Fishlock, 1985). The range of non-antibiotic microbial products is likely to increase as sensitive and rapid assays for a range of biological activities are developed. Indeed, many of the 'failed' antibiotics are being rescreened as potential drugs.

Production of single cell protein

One of the great hopes for fermentation during the 1960s was that large-scale conversion of petrochemical products to microbial biomass could provide a source of protein-rich food and feed. To a large extent, the prospects for 'single-cell protein' (SCP) foundered with the oil crisis of the 1970s. The possible substrate for SCP extended beyond petrochemicals to include methane, methanol, ethanol, sugars, petroleum hydrocarbons, and industrial and agricultural wastes. A variety of micro-organisms has been used for SCP production (Primrose, 1987; Goldberg, 1988). SCP is currently used as high-grade animal feed for intensively reared calves and poultry, and as human food. The Rank Hovis McDougall company in the UK markets a tasteless, odourless, textured fungal protein known as Quorn as a component of 'chicken'- and 'beef'-flavoured pies. Quorn is sold on its merits as a meat substitute.

As yet, commercial production of SCP in the industrialized countries is still very low, because, for the feed market, it has to compete with cheap vegetable or animal products like soyabean and fish meal (Goldberg, 1988). However, SCP production is quite widespread in the USSR where it is seen as a way of converting the natural gas surplus into much needed protein.

At present, the main obstacle to the widespread production of SCP is its high price compared with oil. Higher yielding microbial strains are also needed. Current research efforts using rDNA and other techniques, to improve quantity and quality, may lead to lower costs of production and an economically more viable SCP production (Senez, 1987; Primrose, 1987; Goldberg, 1988).

Biogas production

Biogas production is the anaerobic conversion of carbohydrates, like sugars, and other organic matter into methane and carbon dioxide by mixed populations of micro-organisms. The various microbial populations interact with the end-product of one group serving as the substrate for others. Biogas digesters, sometimes little more than a hole in the ground, are supplied with manure or wastes from livestock, crop residues and so forth. The biogas produced is used as an energy source for domestic purposes while the organic residues (including cellulose and hemicellulose) are used as compost (Senez, 1987). Biogas production both creates energy and treats waste. Biogas digestion is cheap and simple. The only requirements are a closed reactor, substrate (organic matter), water (30-60%), and sometimes some phosphate and/or salts. Overacidification of the digester, the main problem of the process, can be prevented by adding calcium and/or by lowering the water content (Hinsenveld, pers. comm.).

Solid state fermentation

In solid state fermentation (SSF) processes, microbial growth and product formation occur on the surfaces of solid substrates (Mudgett, 1986). Because of the heterogeneity (both physical and chemical) of solid substrates, SSF is not as well characterized as is the fermentation in a liquid medium. It is, however, widely used in developing countries, especially in East Asia. Substrates traditionally fermented in solid state include rice, wheat, millet, barley, corn, soyabean and cassava. The micro-organisms involved include a large number of filamentous fungi and a few bacteria such as species of actinomycetes and *Bacillus* (Mudgett, 1986). SSF includes a number of well-known microbial processes such as composting, wood rotting, mushroom cultivation and the production of animal feed and human food (e.g. mould-ripened cheese, baking, sausage, tempeh, natto, soya sauce, rice wine and kaffir beer).

SSF is an unsophisticated technology. Solid substrates require only the addition of a little water. Low moisture reduces the problem of contamination which so often affects liquid fermentations. Agricultural substrates may, however, require some kind of pretreatment like cracking or surface abrasion (Mudgett, 1986). Technologically, the major problem with SSF is the difficulty of scaling up the process while avoiding heat build-up during fermentation (Mudgett, 1986; Rai *et al.*, 1988).

One interesting, non-traditional application of SSF is the up-grading of lignocellulosic wastes such as crop residues. Crop residues are rich in carbohydrate but low in nutritional value for animal feed. Animals shun them and they are poorly digestible, low in protein and high in heavily lignified fibre. Their digestibility is primarily limited by lignin which acts as a barrier to the breakdown of cellulose and hemicellulose by rumen micro-organisms. Lignin can, however, be degraded in SSF (Mudgett, 1986). Furthermore, the protein content of crop residues may be increased by the presence of the various fungi active in SSF. Without further research, however, costs and technical problems probably prevent lignocellulose SSF from becoming a profitable way of producing animal feed.

Microbial inoculation of plants

Micro-organisms and plants have a close interdependence. Soil, water and air provide a relatively harsh environment for life and cooperation between organisms is vital in the competition for scarce resources. Stripped of their microbial partners, plants may struggle to mobilize nutrients and can become prey to other, disease-causing micro-organisms.

Humans have endeavoured to influence plant-microbial interactions by using microbial inoculants. The technology involves the selection and multiplication of plant-beneficial micro-organisms and applying them to plants, seed or soil. The two main uses of micro-organisms are as biofertilizers for improved plant nutrition and as biological control agents to combat pests, weeds and diseases. Microbial inoculation of plants goes back about 100 years and has been commercialized for the last 50 years. Despite this long history, the global market for microbial fertilizers and pesticides has been, and still is, insignificant compared to those for chemical products (Macdonald, 1989).

Biofertilizers

Many soil micro-organisms enhance nutrient uptake in plants. Those which have a direct beneficial effect on the plant may have considerable potential as biofertilizers. Three groups of plant-beneficial micro-organisms can be distinguished: nitrogen-fixing micro-organisms, mycorrhizal fungi and plant growth-promoting rhizobacteria.

Biological nitrogen fixation (BNF)

Plants cannot use nitrogen from its most abundant source, the atmosphere. Thus harvesting a plant effectively removes nitrogen from the soil. The natural processes for replenishing nitrogen used up by crops are too slow to sustain the productivity demanded by modern agriculture. The shortfall is made up by chemical fertilizers, prepared industrially by taking nitrogen from the atmosphere. Some micro-organisms are also capable of fixing atmospheric nitrogen: the blue-green algae (cyanobacteria), soil bacteria of the genera *Azotobacter, Klebsiella, Bradyrhizobium, Rhizobium*, and Actinomycetes. These nitrogen-fixing bacteria possess nitrogenase, an enzyme which converts atmospheric nitrogen to ammonia. To fuel the process, bacteria need energy. Some, such as the cyanobacteria, use solar energy by combining BNF with photosynthesis. Others, such as *Azotobacter* and *Klebsiella* spp., use organic food. Yet others, notably the rhizobia, enter into symbiotic relationships with plants; the bacteria receive some products of the plants' photosynthesis and, in return, donate ammonia to the plant, which the plant uses for synthesizing proteins (Dixon, 1987; Postgate, 1990).

Ecologically and agriculturally, the most important BNF systems are symbioses. Rhizobia, for instance, as free-living bacteria in soil cannot fix nitrogen. Nitrogen fixation can only occur under anaerobic circumstances, because oxygen degrades the nitrogenase. However, when rhizobia infect the root hairs of suitable host plants (legumes like peas, beans, lupins, clover and *Leucaena*), they induce the formation of nitrogen-fixing nodules in the plant. Each group of leguminous plants has its own species of rhizobium (Postgate, 1990). Another symbiotic association involves the cyanobacterium, *Anabaena azolla*, and the water-fern, *Azolla* spp. The fern, when introduced in paddy fields, can provide rice with nitrogen. *Azolla* can also be intercropped with other plants. Nitrogen fixing is the principal, but not the only reason for increases in rice production resulting from *Azolla*; it can also provide 'green manure'. *Azolla* appears to be most successful in the subtropics (Whitton and Roger, 1989). Free-living blue-green algae can also be directly inoculated on paddy fields in a method called algalization. The effectiveness of algalization is still a matter of some controversy. In general, free-living blue-green algae appear to have less potential in terms of nitrogen fixed than *Rhizobium* or *Azolla* (Whitton and Roger, 1989).

Effective BNF can avoid the use of costly and eventually polluting nitrogen fertilizers. There is much scope for spreading and improving existing technologies without recourse to genetic engineering: agricultural practices can be improved to enhance nitrogen fixation by free-living and symbiotic micro-organisms, more efficient and competitive microbial strains can be selected, higher-yielding association between plant and nitrogen-fixing micro-organisms can be encouraged, and new nitrogen-fixing organisms isolated (Postgate, 1990; UNDP, 1989). Competition between strains of the inoculum and inefficient indigenous strains is the major imponderable in BNF; the mechanisms which determine competitiveness are still poorly understood (Eaglesham, 1989).

Mycorrhizal associations

Mycorrhizal associations are another form of symbiosis - between certain fungi, particularly vesicular-arbuscular (VA)-mycorrhizas, and the roots of vascular plants (e.g. trees, shrubs and certain plants like wheat, sorghum, cassava, soya, tea and coffee.) Unlike rhizobial associations, VA mycorrhizal fungi are non-specific; they can infect a very wide range of host plants; it is assumed that more than 90% of all vascular plants can form mycorrhizal associations. In many circumstances, mycorrhizal infection can greatly increase the rate of uptake of nutrients, particularly phosphorus and nitrogen, from deficient soils. VA mycorrhizas solubilize iron phosphate in low fertility soils. In addition, they may mobilize other trace elements like copper, zinc and possibly iron (Yuthavong and Bhumiratana, 1989; Stribley, 1989). Although mycorrhizal fungi in general are widespread in soils, artificial inoculation could improve host response where the concentration of natural inoculum is suboptimal (which is almost always the case in new plantings), or when the natural inoculum is inferior in its physiological effects on the host. However, large-scale production of pathogen-free mycorrhizal inocula in aseptic culture is not yet possible. Commercial inocula have been produced and marketed to a limited extent in the USA (Stribley, 1989). Furthermore, the use of artificial mycorrhizal inoculants is usually not very effective, largely because our knowledge of mycorrhizal associations is limited. There is little information on the properties and distribution of natural inoculum of VA mycorrhizal fungi. Little is known about genetic variation in VA mycorrhizal fungi, or about the effects of the environment upon natural selection of particular genotypes. Artificial inoculation may be unnecessary if an effective population of native mycorrhizal fungi in the soil can be built up by management practices.

Plant growth-promoting rhizobacteria

Rhizobacteria are bacteria from the rhizosphere (the region around the plant's roots) which can colonize the roots. The term plant growth-promoting rhizobacteria (PGPR) was given to beneficial rhizobacteria in 1978. Most PGPR are fluorescent pseudomonads or strains of *Bacillus subtilis* (Campbell, 1989; Davison, 1988; Kloepper *et al.*, 1989). PGPR can be inoculated on to some crops and can subsequently improve growth. The beneficial effects of PGPR fall into two categories: growth promotion and plant disease suppression. Growth promotion is evidenced by increases in seedling emergence, vigour, seedling weight, root system development and yield (Kloepper *et al.*, 1989). Its mechanisms are not well understood. Some PGPR strains are supposed to increase plant growth by positively interacting with various plant-symbiotic micro-organisms, such as *Rhizobium*, *Bradyrhizobium*, *Frankia* and mycorrhizal fungi. For example, several pseudomonads and *Bacillus* spp. can enhance nodulation (Kloepper *et al.*, 1989). Some PGPR produce plant growth hormones in culture, but there is no evidence that this occurs under normal field conditions or is important in growth promotion (Campbell, 1989). Some PGPR strains do not promote growth directly but appear to control minor root pathogens (see below).

Biological control agents

Certain micro-organisms are natural pesticides, fungicides, bactericides and/or herbicides; over 100 bacteria, fungi and viruses that infect insects have been described. The basic principles of biological control were established about 60 years ago but the approach has been in the shadow of chemical pesticides, fungicides and herbicides. Currently, the most widely used micro-organism in insect control (90% of all microbial insecticides) is *Bacillus thuringiensis* (Bt), an aerobic spore-forming bacterium, which produces a proteinaceous crystal toxic to many insect species including gypsy moths, inchworms, hornworms and cabbage looper (Davison, 1988; Macdonald, 1989). The toxin is very selective and biodegradable. Ingestion of the toxin by a susceptible insect leads to death within a few days. Bt was first isolated in 1911 and has been commercially produced since the 1960s. The only other significant bacterium in insect control is *Bacillus popilliae*. It is used to control the larval stages of the Japanese Beetle in ornamental turf (applied as spore dust). Insect viruses have also been applied to control, for example cotton bollworm, pine caterpillars and spruce sawfly. Fungi can also be used but these control agents are not yet widely commercialized (Macdonald, 1989). Micro-organisms are also, to a limited degree, applied as herbicides. The oldest microbial herbicide is the fungus *Collectrotrichum gloeosporioides* which is used to control weeds in rice and soyabean cultivation. Microbial inoculants are also applied in the control of plant diseases caused by root pathogens. A commercially produced inoculant is the fungus *Trichoderma* which is used against Dutch elm disease (Macdonald, 1989; Campbell, 1989; Baker, 1989). Other promising antagonists are the pseudomonad PGPR which appear to be effective against minor root pathogens. These PGPR may act by producing siderophores (chelating compounds which 'mop up' iron) which reduce the availability of iron for deleterious rhizobacteria or fungi, or by producing toxic compounds like antibiotics and hydrogen cyanide (prussic acid) (Campbell, 1989; Davison, 1988; Kloepper *et al.*, 1989).

A reason for using microbial inoculants in biological control is that micro-organisms have some advantages over chemicals (Table 6; Whitton and Roger, 1989).

Large-scale biological control is promising and is being actively pursued. The production of useful biological control agents, however, is a long process and needs research and development time and money. Most of the microbial inoculants which are now commercial or near-commercial have been worked on, with varying intensity, for about 20 years (Campbell, 1989). A disadvantage of microbial inoculants is that since they are alive, they are more sensitive to environmental changes in the field than chemicals. Furthermore, it is difficult to decide which organisms are worth working on: the organism needs to suit the environment, be competitive with existing organisms and effective as a control agent (Campbell, 1989). One strategy in enhancing biocontrol agents is to combine them with other control methods such as the use of chemicals. Current research on biological control agents is concentrated on minimizing the cost of production of the micro-organisms and searching for effective means for dispersing the micro-organisms (Yuthavong and Bhumiratana, 1989).

Table 6: Comparison of chemicals and microbial inoculants

	Chemicals	*Microbial Inoculants*
Costs/benefits		
R&D	US$20m	US$0.8-1.6m
Market size required for profit	US$40m/year to recoup development costs, therefore limited to major crops	Markets under US$1.6m may be profitable due to low development costs
Toxicological testing	US$ 10m	US$0.5m
Patentability	Well established	Still developing
Lead time	6-7 years	3-5 years
Discovery	Screen 15,000 compounds to identify one product	Rational selection for specific target pests
Efficacy		
Kill	ca. 100%	Usually 90-95%
Speed of kill	Rapid	Can be slow
Spectrum of activity	Generally broad	Generally narrow
Resistance	Often develops	Only one known case
Type of action	Can be both preventive and curative	Generally only curative
Safety		
Operator safety	Chemicals can be hazardous	Low operator risk
Environmental impact	Many examples, e.g. accumulation in food chains	Few examples with use of indigenous micro-organisms
Residues	Interval before harvest often required	Crop can usually be harvested immediately after application

Plant cell and tissue culture

Plant tissue culture

Plant tissue culture has been known for over 30 years. Its basis is the ability of many plant species to regenerate a whole plant from a single cell. Plant cells derived from leaves, roots, anthers, protoplasts or meristems can be grown in a test tube filled with appropriate artificial growth medium under sterile conditions. The growth medium contains essential minerals and plant hormones. After growth in controlled conditions, regenerated plants can be transferred to soil. There are a number of different techniques for plant tissue culture: meristem culture, *in vitro* multiplication, somatic embryogenesis and organogenesis, *in vitro* selection, pollen and anther culture, embryo rescue and somaclonal variation.

In meristem culture, the apical meristem tip (embryonic cells) of a plant is isolated and cultivated *in vitro*. This relatively simple technique can be used to micropropagate cultivars and to produce virus-free plants from infected stock (viruses are usually absent or at greatly reduced concentration in the meristem tip). Meristem culture has been applied to over 50 plant varieties including strawberry, cocoa, grape, lemon, and root and tuber crops like cassava, sweet potato and yam (Sasson, 1989; Dodds, 1989).

In vitro multiplication is a more sophisticated technique by which selected plants can be rapidly propagated (cloned). Leaf or embryo tissue (generally) is used to produce callus, an undifferentiated cluster of plant cells. The callus is divided in single cells or cell clusters each of which can differentiate into embryo-like cells, embryoids or other plant structures, and subsequently regenerate. This technique is particularly useful in multiplying plants that grow slowly or do not yield seed. It has been successfully used, for instance, with banana, plantain, potato, sweet potato, yam and oilpalm.

The tissue masses or calli formed *in vitro* can be transformed into embryoids which strikingly resemble embryos from sexual reproduction. This is somatic embryogenesis. The embryoids can produce plantlets which are grown in greenhouses before being planted outside in nurseries (Primrose, 1987). Using growth hormones, callus can also differentiate into other plant structures like root or leaf structures which can grow into whole plants. This is organogenesis (Nijkamp, pers. comm.).

At the *in vitro* multiplication stage, it is possible to select cells (and therefore plants) with certain characteristics by manipulating the media in which they grow. For instance, cells selected on the basis of improved tolerance to salts, metals, herbicides, or extremes of acidity or alkalinity in culture can give rise to plants exhibiting the same trait(s). This technique has been proven successful in a restricted number of cases (UNDP, 1989).

Pollen and anthers can also be grown in tissue culture. The advantage of this approach is that pollen and anthers are haploid - they contain only one copy of each chromosome. Chromosome duplication can be induced to give rise to fertile, diploid, homozygous (with two identical sets of chromosomes) plants. This method has proved useful in shortening the time to produce homozygous plants which are necessary to achieve the expression of recessive genes (Senez, 1987; UNDP, 1989).

Another interesting aspect of *in vitro* multiplication is the phenomenon of somaclonal variation. When cells from a callus are regenerated in tissue culture, mutations, frequently caused by chromosomal rearrangement (i.e. translocation, insertions, etc.), occur spontaneously. Deletion or amplification of gene sequences also occurs. The resultant plants are often stable and can be used to develop new varieties. Somaclonal variation has been found in all plant species that have been regenerated from tissue culture (Scowcroft, 1989).

Tissue culture is used in germplasm conservation and is an essential tool in biotechnological plant breeding. Once a plant cell's genome has been altered through genetic modification techniques, the transformed cell will have to be propagated in tissue culture.

The most notable technical drawback to tissue culture is that tissues of certain plants have proved difficult to regenerate. Important cereal grains like wheat, oats and barley, which are monocotyledenous are particularly recalcitrant (UNDP, 1989). A second shortcoming is that, as with conventional plant breeding, tissue culture-induced changes affect the entire plant genome: altering one characteristic may induce other changes. The elimination of these traits by back-crossing often takes considerable time and effort. Moreover, the induced change may prove unstable as the plant is propagated in subsequent generations (UNDP, 1989).

Somaclonal variation can be useful as a technique for increasing genetic variation but sometimes this is highly undesirable. Where *in vitro* multiplication is being used with the intention of producing a large number of genetically identical offspring, for instance, somaclonal variation may nevertheless occur. The offspring will differ from the parent plant and may have lost some of the desired characteristics of the parent.

Plant cell culture

Plant cell culture involves the production of secondary metabolites by undifferentiated or differentiated plant cells which are cultured in liquid medium. Plant cell culture is a further development of plant tissue culture and is related to fermentation. Callus tissue in liquid medium, when gently agitated, will shed individual cells. Isolated cells can be transferred to a small flask where, given appropriate conditions, growth and division continues. Some cell line(s) may produce large quantities of secondary metabolites and these can be isolated and further developed (UNDP, 1989). Usually it is necessary to improve the yield of a cell line through physiological or genetic modification of expression of the relevant genes. After growth and production, the desired product will have to be recovered either from the plant vacuole or from the liquid medium. Plant cell culture usually involves growth of undifferentiated cells. Some products, however, can hardly be produced by undifferentiated cells whereas their production also cannot be induced. If these products are specific to a certain part of the plant, (e.g. the roots) then the culturing of differentiated cells (e.g. root culture) might be considered as an alternative.

Plant cells in culture produce a range of compounds from simple sugars to complex molecules that find use in health care (drugs), as food additives (flavours, dyes), and in agriculture (herbicides, insecticides). Already over 50 natural products are being produced in plant cell culture at yields equal to or greater than equivalent crops. These include shikonine, quinine, morphine, nicotine, alkaloids, codeine, cocoa, pyrethrum, thaumatin, jasmine extract, ginseng and digitalis. Cell culture products are typically low-volume, high-cost substances (UNDP, 1989).

The advantage of plant cell culture is that it is free of pest infestation, it is unaffected by climate and weather, its defined growth conditions lead to more stable quality and consistency of product, it ensures production when and as needed, and it greatly reduces land utilization (UNDP, 1989). A political advantage is that it can enhance self-reliance of certain countries by making them independent of imports. However, it is still a developing technology. The fundamental processes by which cultured cells switch on the synthesis of the products are poorly understood. Furthermore, plant cells grow very slowly (much slower than microbial cultures), they are susceptible to destructive physical forces in the reactor vessel ('shear'), and they are prone to infection (UNDP, 1989). It is estimated that the end-product's value has to be greater than US$1000 per kilogram before it could be produced profitably in plant cell culture. Although some plant products are now appearing on the market, the process is not expected to be commercially attractive for many years.

Enzyme technology

Enzymes are proteins that catalyse chemical reactions while remaining unchanged upon the reaction's completion. They are highly specific and biodegradable, and when compared to most chemical processes operate in conditions of mild pH (4-8), low temperature (10-80°C) and at normal atmospheric pressure. Enzyme technology can be divided into four phases (Towalski and Rothman, 1986):

1. *Enzyme extraction.* Extracellular enzymes (enzymes that are secreted by the cells into the environment) from microbial cells can be processed directly from the culture medium. In the case of intracellular enzymes (which catalyse reactions within the cell itself) the cells need to be disrupted to 'free' the enzymes. Surfactants may be needed to dislodge the enzymes from cell fragments. The extracts are usually very dilute and may contain many impurities.

2. *Purification and concentration.* The extract is filtrated or centrifuged, and precipitated.

3. *Stabilization.* The stability of enzymes is improved by the blending of proteins, salts, starch hydrolysates or sugar alcohols.

4. *Preparation.* Enzymes can be prepared either (i) in liquid form as crude extracts or highly purified and stabilized preparations or (ii) in solid form as pure crystals, freeze dried (semi-pure), or immobilized in or on a wide range of carriers.

The use of purified enzymes is over 50 years old. Although more than 2000 enzymes have been characterized, only about 350 are used with a dozen or so accounting for the bulk of those commercially available (Campbell, 1984). All enzymes are obtained from living organisms, usually microbes but sometimes animal organs or plant tissues. The utilization of enzymes has been improved considerably since the advent of immobilization techniques. Immobilization makes the enzyme re-usable and opens up the way to continuous processing. Today, enzymes have five distinct areas of application: (i) as scientific research tools; (ii) in cosmetics; (iii) for diagnostic purposes; (iv) in therapeutic treatment; and (v) for use in industry. One of the important recent applications of enzyme technology is the production of high fructose corn syrup (HFCS), also called isoglucose, from maize (see Chapter 1, Box 1). Enzymes are of particular interest in the production of stereoisomeric compounds. For instance, they have potential application in the production of stereoisomeric pheromones - complex molecules emitted by insects to communicate with other members of their species. Pheromones can be used in integrated pest management to disrupt insect mating (thereby reducing successive populations), or to lure insects to their death by mixing a small amount of insecticide with the pheromone. The commercially available pheromones are, however, made by chemical, not biotechnological processes (Van Brunt, 1987a).

Enzymes may be produced at relatively low cost in virtually unlimited quantities. Their application makes savings possible in fixed capital costs as enzyme-catalysed processes operate under milder conditions of pH, temperature and pressure than their chemical counterparts.

The main disadvantage of enzymes is their instability; they are susceptible to denaturation and inhibition by slight alteration of their physical environment, and are easily inhibited by chemicals (Antebi and Fishlock, 1985; Towalski and Rothman, 1986). To counter this, recombinant DNA technology is being used to increase the stability and activity of enzymes. Another complication with some enzymes is that they require co-factors - thermostable, non-protein molecules that facilitate the transfer of chemical groups. Co-factors are usually very costly to produce and may need to be recovered (Towalski and Rothman, 1986).

Embryo technology

Over the last 20 years, the techniques for recovery, storage and implantation of animal embryos (embryo transfer or ET) have been developed to the stage where they have practical utility in cattle, sheep and goat breeding (Persley, 1990). Following superovulation (induced by follicle-stimulating hormone, FSH), about twelve germ-cells can be flushed from the oviduct of the female animal. The ripe egg cells are fertilized *in vitro*. Subsequently, the embryos are transferred to an incubator. The embryos are harvested and transferred to a carrier mother (surrogate).

The principal benefit of ET is the ability to produce more offspring from an elite female animal than would otherwise be possible. For example, a cow will give birth to about four calves in an average lifetime.

With ET, this could be increased readily to at least 25 calves. This could enhance the rapid expansion of rare genetic stock (e.g. a new breed). ET can also reduce costs of imported cattle because importing animals as embryos is much cheaper. Moreover, by raising these animals in their new home country they are better able to adapt to local environmental conditions (Persley, 1990; Sasson, 1989).

There are additional potential advantages from ET. After *in vitro* fertilization, a single embryo can be cloned through mechanical splitting or transplantation of nuclei, resulting in two or more genetically identical offspring. Another possibility is embryo sexing which could further increase selection intensities, and could permit greater specialization of the beef and milk production functions of a dual purpose population (Persley, 1990). Furthermore, ET is an essential step in the genetic transformation of animals. A more esoteric application is to fuse split embryos to create chimaeras. The genetic material of the two parental embryos remains separate but is dually expressed in the same animal. The resultant offspring is made up of a mosaic of cells, some of which carry genes from one parent and some from the other. The first and most famous example of this method is the geep, a goat/sheep chimaera (*The Economist*, 1988).

ET is still not possible in swine and buffaloes (except for occasional, non-repeated successes). Deep-freeze conservation of swine germ-cells has been impossible; egg cell mortality already occurs at temperatures below 15°C (Sasson, 1989).

Protoplast fusion

A protoplast is a plant cell of which the cell wall has been stripped away by enzyme treatment. Protoplasts can be produced from single plant cells, callus tissue or intact tissues. Protoplasts can be fused with one another (somatic hybridization) by adding an appropriate chemical fusogen or by subjecting the protoplasts to an electric pulse (electrofusion). The fusion initially contains the entire nuclear and cytoplasmic genetic information of the two cells. Subsequently, however, genomes reassort and recombine to produce a wide, and unpredictable variety of gene combinations. If the fused entity or somatic hybrid can reconstruct the cell wall, a hybrid cell is produced which can subsequently proliferate to form a callus and perhaps a whole plant. The plant contains characteristics of both parents. Thus, protoplast fusion allows for the exchange of genetic information between plants without many of the limitation of conventional breeding: protoplast fusion allows the researcher to cross species barriers. Promising results have been obtained with Brassica (male sterility) (UNDP, 1989; Nijkamp, pers. comm.).

The advantage of protoplast fusion is that it enables the transfer of not only the genetic characteristics borne by nuclear DNA but also by cytoplasmic DNA such as that from mitochondria and chloroplasts. This DNA contains genes that code for important characteristics such as photosynthesis, male sterility, resistance to herbicides and diseases, and tolerance to drought (Primrose, 1987; Senez, 1987). Protoplast formation also opens up possibilities for introducing new genetic characteristics without fusion. Wall-less protoplasts allow the passage of isolated DNA from any number of sources.

Protoplast techniques have, however, yielded very few practical applications. Regeneration of the fused product is the main problem. Another important constraint is that certain chromosomes are lost during subsequent cell divisions. It can never be known beforehand if the desired genes will be incorporated in the genome stably; the final, stable genotypic composition is unpredictable (Primrose, 1987; Senez, 1987; UNDP, 1989). Fusion is most successful when the two cells share a single parental source; the further apart the protoplasts are genetically, the greater the problems of survival for the fused product (Primrose, 1987; Senez, 1987; UNDP, 1989). In plant breeding, recombinant DNA technology has been generally preferred over protoplast fusion.

Hybridoma or monoclonal antibody technology

Another kind of fusion forms the basis of monoclonal antibody technology (Box 11; Baker, 1989). An antibody producing mammalian B lymphocyte (from mouse, rat or human) is fused with a type of cancer cell, called myeloma, using chemical fusogens or sendai virus. The B cell brings the ability to produce a particular type of antibody while the cancer cell endows the hybrid with the ability to grow indefinitely in culture. The resultant hybridoma cell will produce large amounts of a single antibody, a monoclonal antibody (MAb) (Baker, 1989; Stribley, 1989). Hybridoma cells can be grown in large fermentors, in immobilized matrices or in the ascites of rodents. MAbs can then be obtained by harvesting the liquid medium from the cultures.

Box 11: What is an antibody?

When a foreign entity (antigen) is introduced into the circulatory system of a higher vertebrate, it stimulates specific white blood cells, the lymphocytes, to produce antibodies. These can combine specifically with the antigen to facilitate its destruction within, or removal from the body. All antibodies have a similar basic structure consisting of two heavy and two light protein chains held together by disulphide bonds. The antigen-binding function of antibodies is contained in the variable regions of these heavy and light chains (the amino acid sequence there varies from one antibody to another). Each antibody will, therefore, have a different amino acid sequence and spatial arrangement. When the immune system of an animal encounters a new antigen, it does not respond to the entire surface of the antigen but to specific antigenic determinants (epitopes) located on it. Thus, a protein antigen may possess several epitopes and would induce the formation of several different antibodies. Following induction of antibody formation, the animal can be bled and the serum fraction obtained. This serum will contain all the different antibodies produced in response to the antigen, and all the other antibodies present in the serum as a result of the animal's previous encounters with other antigens. Such serum is called polyclonal antiserum.

Monoclonal antibodies can be used for diagnostic purposes to identify pathogenic organisms (viruses, bacteria and parasites), tumours, hormones, metabolites, drugs and toxins in plants, body fluids, tissues or foodstuffs. For antigen-MAb binding to be detected and quantified, the monoclonal antibodies must be labelled by linking an enzyme or isotope (radioactive substance) to them. The advantage of MAb tests in diagnostics is their relative speed, accuracy and specificity. As standard reagents, MAbs are undoubtedly superior to polyclonal serum; they are readily available for an indefinite period at standard quality (Persley, 1990; Primrose, 1987; UNDP, 1989). MAbs can also be used in therapeutic treatment, and in protein purification. In this latter case, MAbs will be attached (immobilized) on solid supports. This separation method is especially valuable when the original mixture is voluminous, the variety of substances in that mixture is large, and there is little of the target substance (*The Economist,* 1988; Primrose, 1987; UNDP, 1989).

The first report of hybridoma production was in 1975. MAb production is now widely used commercially, mainly for improved diagnostics in human health care. In crop and animal production, the main use of MAbs will be in the diagnosis of infectious diseases and the detection of hormone levels in female animals (e.g. in the milk progesterone test), to determine if the female animal is 'in heat' (important for timely artificial insemination) or pregnant.

There are no major limitations besides cost in the application of MAbs for diagnostic purposes. Problems that may occur can largely be overcome. Even though MAbs are specific for a given epitope, cross-reactivity may occur; antigens sometimes share common determinants. Selection of non-cross-reactive antibodies is time-consuming (Campbell, 1984). Although cross-reactivity may not be a major problem in diagnostic use, it presents a potential hazard in therapeutic use. (It should be noted, however, that cross-reactivity is much more common with polyclonal antisera.) A second problem is that MAb tests cannot distinguish between antibodies generated by a disease and those generated by a vaccine.

Recombinant DNA technology

Recombinant DNA (rDNA) technology is, in essence, the insertion of an isolated piece of genetic material into the genome of a host cell. The scope of rDNA technology is broad but here we will concentrate on several of the most important practical applications: DNA probes, transformation of micro-organisms, rDNA vaccines, and the transformation of plants and animals.

DNA probes

DNA probes (gene probes) are single-stranded DNA fragments (often tagged with an isotopic, fluorescent or enzyme label) which attach themselves to a piece of DNA or RNA which is complementary to the probe (nucleic acid hybridization). DNA probes, therefore, can be used to look for the presence of specific DNA sequences in organisms or in isolated DNA preparations, which is relevant for the identification, isolation and mapping of genes.

With DNA probes it is possible to identify the presence of micro-organisms, hereditary markers, oncogenes, antibiotic resistance, or any biological character which has its origin in DNA. Tests for infectious diseases in humans, animals and plants are the easiest to produce and, therefore, furthest along in commercialization (Van Brunt and Klausner, 1987). The main advantage of DNA probes in diagnostics is their high specificity and speed; for example, it is possible to detect pathogens which would not be revealed by antibody-based tests (Primrose, 1987).

One particularly important use of DNA probes is in restriction fragment length polymorphisms (RFLP). The DNA sequences of individuals within a species differ in minor ways; DNA is polymorphic. Sometimes these differences produce heritable characteristics such as eye colour or genetic disease. More often than not, however, the differences are 'silent'. They can be revealed by highly specific enzymes called restriction enzymes which cut DNA molecules in places where there is a specific sequence of bases (Persley, 1990; Primrose, 1987). Treating DNA from different individuals with restriction enzymes produces different sets of DNA fragments. These can be resolved by size using a technique known as gel electrophoresis. Following electrophoresis, DNA probes are used to reveal on which fragment a particular DNA sequence occurs. The result is a kind of 'bar-code' specific for each individual (West *et al.*, 1989).

RFLP technology has been applied in DNA fingerprinting (determination of identity, relatedness, etc.) in forensic investigations, paternity suits and immigration procedures. In plants and animals it is also used to map chromosomal regions involved in traits of economic value (trait mapping). Furthermore, RFLP analysis is used in plant breeding to identify offspring that carry all or most of the desired genes (West *et al.*, 1989). DNA probes are usually produced by chemical synthesis with oligonucleotide synthesizers ('gene machines'). Once a probe is available, it can be used in a techniques known as PCR (polymerase chain reaction) to clone genes. PCR extends the sensitivity of DNA probe-based diagnostics considerably (Nijkamp, pers. comm.).

One major problem with DNA probes is labelling. Until 1987, all probes had isotopic labelling. Isotopes are potentially dangerous, hard-to-handle and their use is time-consuming. Fluorescent substances have subsequently been developed but these also require dedicated instrumentation to read the results. Enzyme-based assays which produce colorimetric end-points can be used to circumvent this problem but these are generally less sensitive than isotopic and fluorescent methods. Various enhancement techniques are being developed to increase the sensitivity of enzyme-based assays (Van Brunt and Klausner, 1987).

Transformation of micro-organisms

Micro-organisms such as bacteria, yeast and fungi are the traditional workhorses of the fermentation industry. Recombinant DNA technology has conferred the possibility to change the characteristics of micro-organisms specifically. The application of rDNA technology to micro-organisms involves the identification, isolation, characterization, transfer, cloning and expression of genes (Dixon, 1987; *The Economist*, 1988; Fairtlough, 1986; Persley, 1990; Primrose, 1987; UNDP, 1989).

Each gene contains the information a cell requires to produce a specific protein. If a protein has been isolated and partially sequenced, corresponding DNA can be used to 'fish out' the corresponding gene. To isolate the DNA, the cells which contain it must be grown and then disrupted through physical means (homogenization) or by using biochemicals such as enzymes or detergents to release the DNA. Then the DNA is purified by centrifugation and electrophoresis.

Once isolated and purified, the DNA can be mapped (see RFLP technology) and sequenced. Subsequently, the required part of the gene can be trimmed and manipulated in the test-tube to prepare it in a form suitable for transfer to a new host. Vectors are commonly used to facilitate gene transfer. A vector can be a plasmid (circle of extrachromosomal DNA), virus, or transposon (segment of DNA which can move from place to place within a genome). In all cases, the vector is designed to help the foreign gene replicate within the new host. The vector will contain certain control sequences which ensure that the gene is transcribed into RNA and translated into a protein correctly. The vector may also contain additional sequences which allow researchers to detect which of the host organisms contain the gene. For instance, genes on the vector which encode for anti-biotic resistance will help the host cell grow on media containing antibiotics: cells without the vector (and without the new gene) will not grow.

The hybrid vectors are incubated with the cells targeted for transformation, the host cells, in an appropriate chemical solution that promotes the entrance of the hybrid vectors into these cells. After entering the target host cell, the vector either becomes incorporated into its genome or remains as a discrete entity (plasmid or episome), which replicates each time the host cell divides (to maintain the acquired genetic information in the transformed cells). The frequency of cell transformation varies with the type of vector, but is usually low. The transformed cells are plated out on agar-containing medium in Petri dishes and left to form individual colonies.

An alternative to the use of vectors for gene transfer is electroporation whereby pores in the cell wall which allow the ingress of DNA can be induced by the application of electric fields to a culture. A similar result can be achieved through laser beams. Another recent technique is the use of microprojectiles; small bullets which are coated with DNA and either fired or accelerated through an electrical field at the host cell. However, unless vectors are used, these techniques lead to the non-specific integration of the gene at random sites in the host cell or organelle chromosome.

Subsequently, the cells containing the new gene (identified by antibiotic resistance or some other marker) can be isolated and grown in large quantities and, if the gene is present on a plasmid, it can be amplified within each cell, thus producing multiple copies. Finally, the gene may be expressed in the transformed cell to produce the desired protein.

Using rDNA technology, one can, therefore, change the characteristics of micro-organisms by, for instance, increasing or decreasing the production of certain metabolic products or allowing the synthesis of completely new products. Bacteria and yeasts have been used to produce a range of products: hormones like somatotropin (growth hormone), somatostatin and insulin which are used for therapeutic purposes; lymphokines (immune system controlling agents) like interferon and interleukin-2 which are also used for therapeutic purposes; and antigens which are used as vaccines. High-yielding bacteria for use in fermentation processes, and plant-beneficial bacteria with new combinations of desirable traits not present in nature have also been produced (Davison, 1988; Fairtlough, 1986).

The case of bovine somatotropin (BST), a protein hormone produced by the pituitary gland, shows what can be done. BST enhances growth, lactation, and influences the balance of carbohydrate and fat metabolism. When injected into cows, it increases milk production by 15-30%, partly by boosting the cow's appetite, and partly by diverting a larger proportion of her food from ordinary metabolism into the production of milk. In the late 1970s, biotechnologists inserted the gene for BST production into bacterial cells (*Bacillus subtilis*) and could produce BST on a large scale (MacKenzie, 1988). Chicken growth hormone and porcine somatotropin (PST) have also been produced.

The 'Ice-minus' bacteria are another example of the genetic engineer's art. Plant pathogenic strains of the bacterium *Pseudomonas syringae* cause frost damage to plants by producing an ice nucleation protein that initiates ice crystalization. In the 'Ice-minus' strain, the ice nucleation gene has been deleted. If this strain is sprayed on the crop it will successfully compete with the 'Ice-plus' strain, thus preventing frost damage.

rDNA vaccines

Vaccines must achieve active and long-lasting immunity (humoral and/or cell-mediated) to prevent the development of disease upon exposure to the corresponding pathogen. However, vaccines themselves must not be pathogenic. It is this separation of immunogenicity from pathogenicity which is the crucial aspect of vaccine development. Conventional vaccines can be put in three main groups. The first contains live, attenuated organisms: by breeding the original pathogen in the laboratory for numerous generations, the organism may lose its virulence while retaining immunogenicity. However, live, attenuated vaccines can revert to virulence. The second category contains inactivated organisms. Inactivation by heat or chemical treatment destroys pathogenicity. The third group contains subunit vaccines. Specific antigens are extracted from the pathogen and purified to serve as the vaccinating substance. The drawback of both inactivated and conventional subunit vaccines, is that the vaccines do not persist in the body and several doses may be required to produce durable immunity.

The emergence of rDNA technology has led to the development of a new range of vaccines. In deletion mutant vaccine, for instance, the genes coding for the virulence (components of the organisms responsible for their pathological characteristics in the host) are removed. The mutated live organism will be non-virulent, but still capable of eliciting an immune response.

The range of potential subunit vaccines has also been extended by recombinant DNA technology. Once the protein antigens responsible for immunogenicity have been identified, researchers can track down their genes and express them in a foreign host, usually a bacterial, yeast or mammalian cell. The isolated protein or peptide can be used to elicit an appropriate immune response (Muscoplat, 1989; Tartaglia and Paoletti, 1988). It is also possible to attach a DNA sequence coding for an immunogenic protein to the gene for an immunogenic protein of another pathogenic organism. Expression of the hybrid gene can yield a hybrid antigen which elicits an immune response to both pathogenic organisms (multivalent vaccine).

The third category of recombinant DNA vaccines is the carrier vaccine. Genes coding for immunogenic antigens can be introduced into the host organism, which is a known non-pathogen, in such a way that the organism expresses the gene on its surface. The vaccinia virus, the fowlpox virus and salmonella bacteria have all been used (Muscoplat, 1989; de Graaf, pers. comm.). When administered as a vaccine, the host organism both expresses the inserted gene and persists in the body. It is, therefore, a live vaccine but one which cannot revert to virulence. Furthermore, it is possible to insert genes coding for antigens of more than one pathogen to produce a multivalent vaccine (Tartaglia and Paoletti, 1988).

Recombinant DNA vaccines are more stable and safer, and can be developed against diseases which cannot be prevented through conventional vaccines. Their development is, however, more costly.

The first rDNA vaccine was developed in 1982; a subunit vaccine against neonatal diarrhoea in piglets and calves (caused by enterotoxigenic *Escherichia coli*) (de Graaf, pers. comm.). Several rDNA vaccines have been developed so far; e.g. rDNA vaccines against bacterial coccidiosis (avian), bacterial brucellosis (bovine), pseudorabies virus (porcine), rinderpest virus (bovine), foot-and-mouth disease virus (bovine), Newcastle virus (avian), papilloma virus (bovine) and feline leukaemia virus (Muscoplat, 1989; Sasson, 1989; Van Brunt, 1987b). Very few of these candidate vaccines are currently commercially produced. So far, no rDNA vaccines have been developed against parasitic diseases.

A disadvantage of recombinant subunit vaccines is that, like conventional, killed vaccines, they are inert and cannot be amplified in the body (Tartaglia and Paoletti, 1988). They must be administered with adjuvants which boost their effects. However, the adjuvants currently used sometimes cause toxic side-effects. A better understanding of adjuvant activity is needed for progress here. Another potential problem with recombinant DNA vaccines is that the expression of the foreign genes in a host cell may not produce the proper post-translational processing (i.e. proteolytic processing and/or glycosylation) which may be important for an optimal immune response (Tartaglia and Paoletti, 1988).

Transformation of plants and animals

The techniques of rDNA technology to transform plants or animals are basically the same as the ones used to transform micro-organisms. Locating a gene in animal or plant cells is more difficult than in micro-organisms largely because their DNA is not only more extensive but also contains long non-coding regions. Specific genes may be localized through DNA mapping, transposon tagging or antisense-RNA approaches. Transposon tagging involves the introduction of a mobile genetic element, a transposon, in the genome. If the plant is subsequently seen to lose a particular function, it can be assumed that this is due to the insertion of the transposon within the corresponding gene. The transposon (and the gene) can then be traced by a transposon (DNA) probe. Antisense-RNA could be used to characterize a certain gene. If a gene is isolated but its purpose is unknown, it is possible to synthesize antisense-RNA which could block the expression of the desired gene. Comparison with untreated plants might reveal the gene's function (Nijkamp, pers. comm.). These different methods enlarge the possibilities of researchers to identify and select desired traits.

With plants only a few vectors are available. The most important is the Ti plasmid, tumour-inducing factor, which is found in the bacterium *Agrobacterium tumefaciens* (the causative organism of crown-gall tumours that often affect wounded plants). The Ti plasmid can insert itself in the plant chromosome and is a useful vector. The Ti plasmid has been most extensively used as a vector for gene insertion into dicotyledenous plants although it can under certain circumstances be used for monocotyledenous plants as well (Dixon, 1987; *The Economist,* 1988; Sasson, 1989; UNDP, 1989). Another vector system which is used for introducing foreign genes into plants is the Ri plasmid (root-inducing factor) of the bacterium *Agrobacterium rhizogenes* which is also taken up by the plant's chromosome (Sasson, 1989; UNDP, 1989). Several viruses that cause plant diseases are promising vectors, among them the cauliflower mosaic virus that infects plants such as cabbage and mustard, the gemini viruses that infect many crops including monocotyledenous plants and that are transmitted by insects, and the RNA viruses, especially the brome mosaic virus (UNDP, 1989).

As with micro-organisms, DNA sequences can be introduced into plants' host cells through direct DNA uptake: microprojectiles and electroporation. Another direct technique is the use of micro-injection. This requires elaborate apparatus consisting of a microscope, a micromanipulator, and an extremely fine glass pipette. Methods of direct DNA uptake offer an alternative route to the genetic transformation of recalcitrant plants, and animals which are not (easily) infected by *Agrobacterium tumefaciens.*

In animals, transformation of the host cell (pronucleus stage or early embryonic stage) is mainly achieved by micro-injection. Electroporation and virus infection (papova-viruses and retro-viruses) have also been applied. The transformed cell must be regenerated via embryo transfer techniques.

Recombinant DNA technology in plant and animal breeding makes it possible to cross-breed species that are too far apart for normal sexual reproduction and to incorporate specific characteristics. Furthermore, biotechnological plant breeding can reduce the time-span needed to develop a new plant variety. It takes about 10-20 years to develop a new plant variety through conventional plant breeding while rDNA technology achieves the same result in 5-10 years (Nijkamp, pers. comm.). Genetic engineering of plants and animals is, however, at an early stage of development. More research is needed on the mechanisms regulating the expression of genes transferred to host cells, as well as on the relation between the desired characteristics and the corresponding gene(s) or gene clusters.

In plants, research on gene transfer began in the early 1970s and resulted, in the beginning of the 1980s, in the development of transfer vectors. The first foreign gene transfer into a plant occurred in 1983 when a bacterial gene coding for antibiotic resistance was transferred to petunia. Since then, plant biotechnologists have successfully transferred single genes controlling agronomically important traits such as resistance to viruses, insects and herbicides, and tolerance to toxic minerals and earliness (Sasson, 1989). More than twenty species, including cabbage, carrot, cotton, lettuce, rape seed oil, peas, petunia, potato, sugarbeet, tobacco and tomato, have been genetically modified. Since 1988, it has been possible to transfer genes coding for antisense-RNA. By blocking the expression of certain genes, it has proved possible to change the colour of flowers (e.g. petunia) and to reduce post-harvest biodegradation (e.g. tomato ripening).

The first incorporation and expression of a foreign gene into an animal genome, was completed in 1980. Gene transfer in animals was pioneered in the mouse: transfer of a gene coding for rat growth hormone produced a transgenic mouse which grew to a much greater size than normal. Biotechnologists have been able to transfer genes coding for a growth hormone in farm animals (swine, sheep and cattle). Since 1982, hundreds of different genes, human and animal, have been inserted into a variety of mammals, birds and fish (*The Economist*, 1988; Persley, 1990). At the moment, however, there are very few commercial applications of genetic engineering in animals.

Much basic research is needed on the biochemistry of plant systems so the genes capable of improving plants may be identified. The mechanisms whereby the foreign gene is taken up by, and becomes part of, the genome are still poorly understood. More often than not, an empirical approach is necessary to find out whether a particular type of cell will be transformed (*The Economist*, 1988; Primrose, 1987; UNDP, 1989). At this point, only a restricted number of dicotyledenous plants have been successfully transformed and then only with single gene characteristics. Many interesting traits, such as crop yield, product quality, drought tolerance, and capacity of nitrogen fixation and photosynthesis, depend on the coherent and complementary actions of many genes (polygenic traits). Furthermore, several genes of importance are not located on the nuclear chromosome (e.g. genes for photosynthesis). Very little is known on how to modify the genes of cell organelles. Although the frequency of transformation with *Agrobacterium* is high (about 30% success), that of other techniques is much lower (Nijkamp, pers. comm.). And the transformation of monocotyledenous plants is more difficult still since they are not (easily) infected by *Agrobacterium*. Finally, there are barriers to overcome in the regeneration and propagation of transformed cells before useful plants can be produced more readily. To circumvent the regeneration problem, much research is being focused upon delivering genes into pollen grains, seeds, or seedlings, all of which can hopefully develop normally into mature plants (UNDP, 1989).

Developing transformed animals is still a time-consuming procedure. For example, from the germ cells collected from a donor mother mouse, 85% prove suitable for injection after *in vitro* fertilization; of those injected, only 70% survive the procedure. Just 10% of the injected embryos that are returned to the foster mother result in live births, and only 20% of the offspring will be transformed. This results in a transformation frequency of 1-2% (*The Economist*, 1988). The mechanism whereby the foreign DNA is taken up and incorporated into the embryo's genome is unknown. It is known that a sort of repair mechanism occurs in a germ cell after fertilization as a result of the damage caused by sperm as it enters it. One assumption is that foreign DNA is picked up sometime during this repair mechanism (*The Economist,* 1988).

Working with large domestic animals is much more difficult than with mice; they do not produce as many germ cells, 'reimplantation of transformed embryos is more difficult because many animals will not give birth to more than two offspring, and the pronuclei or nuclei are often impossible to detect (Primrose, 1987). Furthermore, most functions in animals are polygenic traits, controlled by a multitude of genes. The ability to manipulate multiple genes is in animals, just like in plants, not yet possible (*The Economist,* 1988; UNDP, 1989).

Resource persons

Prof dr F.K. de Graaf, Department of Molecular and Cellular Microbiology, Faculty of Biology, Vrije Universiteit Amsterdam, The Netherlands.

Ir M. Hinsenveld, Department of Civil and Environmental Engineering, University of Cincinnati, Cincinnati, Ohio, USA.

Dr J. van Minnen, Department of Histology, Faculty of Biology, Vrije Universiteit Amsterdam, The Netherlands.

Prof dr H.J.J. Nijkamp, Department of Plant Genetics, Faculty of Biology, Vrije Universiteit Amsterdam, The Netherlands.

Prof dr A.H. Stouthamer, Department of Microbial Physiology, Faculty of Biology, Vrije Universiteit Amsterdam, The Netherlands.

References

Ahmed, I.F.T. (1988) The Bio-revolution in Agriculture: key to poverty alleviation in the Third World. ILO Studies, *International Labour Review*, **127**(1).

Anbar, M. (1986) The 'Bridge Scientist' and his role In: *Interdisciplinary Analysis and Research: Theory and practice of problem-focussed research and development*, D.E. Chubin, A.L. Porter, F.A. Rossini, and T. Connolly (eds). Lomond Publications, Maryland, USA, 155-163.

Antebi, E. and Fishlock, D. (1985) *Biotechnology: Strategies for Life*. MIT Press, Cambridge, Massachusetts, London, England.

Baker, R. (1989) Improved *Trichoderma* spp. for promoting crop productivity. *Trends in Biotechnology*, **7**, February, 34-38.

Biggs, S.D. (1989) Resource-poor farmer participation in research: a synthesis of experiences from nine national agricultural research systems. *Special Series on the Organization and Management of On-Farm Client-Oriented Research* (OFCOR) - Comparative Study Paper No.3, International Service for National Agricultural Research (ISNAR).

Blume, S.S. (1987) The theoretical significance of co-operative research In: *The Social Direction of the Public Sciences: Causes and consequences of cooperation between scientists and non-scientific groups*, S. Blume, J. Bunders, L. Leydesdorff and R. Whitley (eds). D.Reidel Publishing Company, 3-38.

Blume, S., Bunders, J.F.G., Leydesdorff, L., and Whitley, R. (eds) (1987) *The Social Direction of the Public Sciences: Causes and consequences of cooperation between scientists and non-scientific groups*. D. Reidel Publishing Company.

Boon, L.J. and Bunders, J.F.G. (1987) What can plant biotechnologists learn from the Green Revolution In: *Proc. 4th European Congress on Biotechnology*, **4**, O.M. Neyssel, R.R. van der Meer and K.Ch.A.M. Luyben (eds). Elsevier, Amsterdam, The Netherlands, 439-448.

Broerse, J.E.W. (1990) Biotechnology: a challenge or a threat? In: *Research and Development Cooperation: The Role of the Netherlands*, C. Schweigman and U.T. Bosma (eds). Royal Tropical Institute, Amsterdam, The Netherlands, 123-140.

Broerse, J.E.W. (1990-a) Country case study Pakistan. Supplement to Bunders *et al.* (1990). Department of Biology and Society, Vrije Universiteit Amsterdam, The Netherlands.

Brouwer, H. et al. (1991) Country case study Kenya; an inventory study. Department of Biology and Society, Vrije Universiteit Amsterdam, The Netherlands.

Bukman, P. (1989) The government role in biotechnology and development cooperation. *Trends in Biotechnology*, **7**(1), S27-S31.

Bunders, J.F.G. (1987) The practical management of scientists' actions: the influence of patterns of knowledge development in biology on cooperations between university biologists and non-scientists In: *The Social Direction of the Public Sciences: Causes and consequences of cooperation between scientists and non-scientific groups*, S. Blume, J. Bunders, L. Leydesdorff and R. Whitley (eds). D. Reidel Publishing Company, 39-72.

Bunders, J.F.G. and de Bruin, J. (1985) Hoe plantenbiotechnologisch onderzoek wordt afgestemd op de belangen van het bedrijfsleven. *Wetenschap & Samenleving 4, 'En de boer...ploegt hij voort?* Over de maatschappleijke aspekten van plantenbiotechnologie', 18-25 (only available in Dutch).

Bunders, J.F.G. and Leydesdorff, L. (1987) The causes and consequences of collaborations between scientists and non-scientific groups In: *The Social Direction of the Public Sciences: Causes and consequences of cooperation between scientists and non-scientific groups*, S. Blume, J. Bunders, L. Leydesdorff and R. Whitley (eds). D. Reidel Publishing Company, 331-347.

Bunders, J.F.G. (1988) Appropriate biotechnology for sustainable agriculture in developing countries. *Trends in Biotechnology*, 6(8), 173-180.

Bunders, J.F.G. *et al.* (1990) *Biotechnology for Small-Scale Farmers in Developing Countries: Analysis and assessment procedures*. VU University Press, Amsterdam.

Campbell, A.M. (1984) *Monocolonal Antibody Technology*. Elsevier, New York, USA.

Campbell, R. (1989) The Use of Microbial Inoculants in the Biological Control of Plant Diseases In: *Microbial Inoculation of Crop Plants*, R. Campbell and R.M. Macdonald (eds). Special publications of the Society for General Microbiology, 25, IRL Press, Oxford, England, 67-77.

CARDI (1978) A proposal for the establishment of a yam seed propagation scheme in Barabados.

CARDI (1979) A proposal for the establishment of a yam seed propagation scheme in the Commonwealth Caribbean as the development phase of Yam Virus Research. Scheme R 3218.

CARDI (1981) Virus-tested yam tuber multiplication project. A project funded by EDF through the Caribbean Development Bank, Annual Report 1981.

CARDI (1982) Virus-tested yam tuber multiplication project. A project funded by EDF through the Caribbean Development Bank, Annual Report 1982.

CARDI (1983) 1983 Annual Report Barbados Unit.

CARDI (1984) Virus-tested yam tuber multiplication project. A project funded by EDF through Caribbean Development Bank, Final Report 1980-84.

CARDI (1985) 1985 Annual Report Barbados Unit.

CARDI (1986) 1986 Annual Report Barbados Unit.

CARDI (1987) 1987 Annual Report Barbados Unit.

CARDI (1989) Annual Technical Report 1988/89 Barbados Unit.

Chambers, R. (1985) *Managing Rural Development: Ideas and experience from East Africa* . Kumarian Press, Connecticut, USA.

Chambers, R. (1990) Participatory appraisal for rural development: practical approaches and methods Notes for participants in the AKRSP Training Workshop in the Surendrangar Programme Area (India), 16-21 July 1990.

Chambers, R. and Ghildyal, B.P. (1985) Agricultural research for resource-poor farmers: the Farmer-First-and-Last Model. Discussion Paper, Institute for Development Studies Publications, University of Sussex, Brighton, England.

Clark, B.R. (1972) *The organizational saga in higher education.* ASQ 17.

Clark, N. and Juma, C. (1991) *Biotechnology for Sustainable Development: Policy Options for Developing Countries.* International Federation of Institutes for Advanced Studies (IFIAS), Research series no.10, ACTS Press, Nairobi, Kenya.

Collinson, M. (1988) FSR in evolution; past and future. Plenary Keynote Address to the 8th Annual Farming Systems Research/Extension Symposium. October 9-12 1988, Winrock International, University of Arkansas. Arkansas USA.

Cramer, J., Eyerman, R. and Jamison, A. (1987) The knowledge interests of the environmental movement and its potential for influencing the development of science In: *The Social Direction of the Public Sciences: Causes and consequences of cooperation between scientists and non-scientific groups,* S. Blume, J. Bunders, L. Leydesdorff and R. Whitley (eds). D. Reidel Publishing Company, 89-115.

CIMMYT Economics Staff (1984) The farming systems perspective and farmer participation in the development of Appropriate Technology. In: *Agricultural Development in the Third World,* Carl. K Eicher and John. M. Staatz, (eds).

Davison, J. (1988) Plant-beneficial Bacteria. *Bio/Technology,* **6,** March 1988, 282-286.

De Zago, M.B. (1986) Interdiscipline: search and discovery - systematization, application and transfer. *Impact of Science on Society,* **28**(2), 127-137.

Dixon, B. (1987) The gene revolution. *The Unesco Courier,* March, 13-16.

DGIS (1989) Biotechnology and Development Cooperation: Inventory of the biotechnology policy and activities of a number of Donor Countries and Organizations, UN Agencies, Development Banks, and CGIAR. Report of the Netherlands Directorate General for International Cooperation.

Dodds, J.H. (1989) Tissue culture techniques for germplasm improvement and distribution In: *Strengthening Collaboration in Biotechnology: International agricultural research and the private sector.* Proceedings of a Conference held April 17-21, 1988 in Rosslyn, USA. Bureau for Science and Technology, Office of Agriculture, Agency for International Development (AID), Washington, USA, 109-128.

DPO/OT (1988) Research and Technology. Section for Research and Technology of the NGO, Education and Research Programmes Department (DPO/OT) of the Directorate General for International Cooperation, Ministry of Foreign Affairs, The Hague, The Netherlands.

Drucker, P. (1985) *Innovation and Entrepreneurship: Practice and principles.* Pan Books, London, UK.

Eaglesham, A.R.J. (1989) Global importance of rhizobium as an inoculant. In: *Microbial Inoculation of Crop Plants,* R. Campbell and R.M. Macdonald (eds). Special publications of the Society for General Microbiology, **25,** IRL Press, Oxford, England, 29-48.

The Economist (1988) The genetic alternative: A survey of biotechnology. *The Economist,* 30 April, 58.

Elz, D. (ed.) (1984) The planning and management of agricultural research. A World Bank and ISNAR Symposium, The World Bank, Washington, USA.

Fairtlough, G.H. (1986) Genetic engineering: Problems and opportunities. In: *The Biotechnological Challenge*, S. Jacobssen, A. Jamison and H. Rothman (eds). Cambridge University Press, 12-36.

Farrington, J. and Martin, A.M. (1987) Farmer participatory research: a review of concepts and practices. *ODI Discussion Paper*, **19**, London, UK.

Farrington, J. and Martin, A.M. (1988) Farmer participatory research: a review of concepts and recent fieldwork. *Agric. Admin. & Extension*, **25**, 247-264.

Gebremeskel, T. and Oyewole, D.B. (1987) *Yam in Africa and the World: Trends of vital statistics 1965-1984*. International Institute for Tropical Agriculture (IITA) Socioeconomic Unit, Ibadan, Nigeria.

Goldberg, I. (1988) Future prospects of genetically engineered single cell proteins. *Trends in Biotechnology*, **6**, February , 32-34.

Greenshields, R. and Rothman, H. (1986) Biotechnology and fermentation technology. In: *The Biotechnological Challenge*, S. Jacobssen, A. Jamison and H. Rothman (eds). Cambridge University Press, 77-95.

Hawes, F. (1980) The cultural factor in the transfer of technology to developing countries. *The Bridge*, **5**(3), 17-19, 40-41.

Horton, D. and Prain, G. (1989) Beyond FSR: New challenges for social scientists in agricultural R&D. *Quarterly Journal of International Agriculture*, **28**(3/4), 301-314.

ILEIA* (1988) Participative technology development. *ILEIA Newsletter*, **4**(3), October.

ILEIA (1989) Participatory technology development in sustainable agriculture. *Proceedings of ILEIA Workshop on 'Operational Approaches for Participative Technology Development in Sustainable Agriculture'*, Leusden, The Netherlands.

ILEIA (1990) Participatory technology development in sustainable agriculture. *Proceedings of a Workshop Information Centre for Low-External-Input Agriculture*.

Joffe, S. (1986) Biotechnology and Third World Agriculture: a literature review and report. Institute of Development Studies, Sussex, UK.

Junne, G.C.A. (1987) Bottlenecks in the diffusion of biotechnology form the research system into developing countries' agriculture. In: *Proc. 4th European Congress on Biotechnology*, **4**, O.M. Neyssel, R.R. van der Meer and K.Ch.A.M. Luyben (eds). Elsevier, Amsterdam, The Netherlands, 449-458.

Kaimowitz, D. and Merrill-Sands, D. (1989) *Making the link between agricultural research and technology users*. Discussion paper for the Workshop International Service for National Agricultural Research.

Kanter, R.M. (1983) *The change master: Innovations for productivity in the American corporation*. Simon & Schuster, New York, USA.

Kenney, M. (1986) *Biotechnology: The university-industrial complex*. Yale University Press, New Haven, USA.

---

* Information Centre for Low-External-Input Agriculture

Klein, J.T. (1986) The broad scope of interdisciplinarity In: *Interdisciplinary Analysis and Research: Theory and practice of problem-focussed research and development*, D.E. Chubin, A.L. Porter, F.A. Rossini, and T. Connolly (eds). Lomond Publications, Maryland, USA, 409-424.

Kloepper, J.W., Lifshitz, R. and Zablotowicz, R.M. (1989) Free-living bacterial inocula for enhancing crop productivity. *Trends in Biotechnology*, **7**, February, 39-44.

Lessem, R. (1986) *Intrapreneurship: How to be an enterprising individual in a successful business*. Wildwood House, Hampshire, UK.

Leydesdorff, L. and Van den Besselaar, P. (1987) What we have learned from the Amsterdam Science Shop. In: *The Social Direction of the Public Sciences: Causes and consequences of cooperation between scientists and non-scientific groups*, S. Blume, J. Bunders, L. Leydesdorff and R. Whitley (eds). D. Reidel Publishing Company, 135-160.

Lipton, M. and Longhurst, R. (1989) *New Seeds and Poor People*. Unwin Hyman, London, UK.

Macdonald, R.M. (1989) An overview of crop inoculation. In: *Microbial Inoculation of Crop Plants*, R. Campbell and R.M. Macdonald (eds). Special publications of the Society for General Microbiology, **25**, IRL Press, Oxford, England, 1-9.

MacKenzie, D. (1988) Science milked for all it's worth. *New Scientist*, 24 March, 28-29.

Mantell, S.H. and Haque, S.Q. (1979) Disease-free yams: Their production, maintenance and performance. *Yam Virus Project Bulletin*, (2), CARDI, July.

Mantell, S.H., Haque, S.Q. and Whitehall, A.P. (1979) A rapid propagation system for yams. *Yam Virus Project Bulletin*, (1), CARDI, July.

Mintzberg, H. (1979) *The Structuring of Organizations*. The Theory of Management Policy Series, Prentice-Hall.

Mowery, D. and Rosenberg, N. (1979) The influence of market demand on innovation: a critical review of some recent empirical studies. *Research Policy*, **8**, 102-153.

Mudgett, R.E. (1986) Solid state fermentation In: *Manual of Industrial Microbiology and Biotechnology*, A.L. Demain and N.A. Solomon (eds), Washington 66-83.

Muscoplat, C.C. (1989) Commercialization and research perspectives for vaccines. In: *Strengthening Collaboration in Biotechnology: International agricultural research and the private sector*. Proceedings of a Conference held 17-21 April, 1988 in Rosslyn, USA. Bureau for Science and Technology, Office of Agriculture, Agency for International Development (AID), Washington, USA, 141-149.

Okereke, G.U. (1988) *Technology and employment programme: biotechnology to combat malnutrition in Nigeria*. World Employment Programme Research, Working Paper, ILO, Geneva.

OECD (1986) *Recombinant DNA safety considerations: Agricultural and environmental applications of organisms derived by recombinant DNA techniques*. OECD Report, Organization for Economic Cooperation and Development, Paris, France.

OTA (1984) *Commercial biotechnology: An international analysis.* US Congress Office of Technology Assessment, OTA-BA-81, Washington D.C.

Persley, G.J. (1990) *Beyond Mendel's Garden: Biotechnology in the service of world agriculture.* Biotechnology in Agriculture series, 1, CAB International, Oxon, UK.

Pinchot, G. (1985) *Intrapreneuring: Why you don't have to leave the corporation to become an entrepreneur.* Perennial Library, Harper & Row, New York, USA.

Postgate, J. (1990) Fixing the nitrogen fixers. *New Scientist,* 3 February, 57-61.

Primrose, S.B. (1987) *Modern Biotechnology.* Blackwell Scientific Publications.

Rai, S.N., Singh, K., Gupta, B.N. and Walli, T.K. (1988) Microbial conversion of crop residues with reference to its energy utilization by ruminants: An overview In: *Fibrous Crop Residues as Animal Feed,* K. Singh and J.B. Schiere (eds). Proceesing of an international workshop held 27-28 October, Bangalore, India.

Rhoades, R. E. (1984) *Breaking New Ground: Agricultural anthropology.* International Potato Centre, Lima, Peru.

Richards, P. (1986) What's wrong with farming systems research. *Paper for the Conference of the Development Studies Association,* University of East Anglia, 15-16 September.

Roobeek, A. (1987) *Biotechnology: A challenge full of promise and pitfalls.* Brochure of the Socialist Group, European Parliament.

Roobeek, A.J.M. (1990) *Beyond the Technology Race: An analysis of technology policy in seven industrial countries.* Elsevier, Amsterdam.

Ruivenkamp, G. (1989) De invoering van biotechnologie in de agro-industrile produktieketen: de overgang naar een nieuwe arbeidsorganisatie. PhD thesis (only available in Dutch).

Sasson, A. (1989) *Biotechnologies and developing countries: present and future.* Information document and basis for a keynote address to the FAO/CTA Symposium 'Plant Biotechnology for Developing Countries', Luxembourg, 26-30 June.

Schmitt, R.W. (1985) Continuity and change in the US research system. School of Public Policy, George Washington University, Occ. Papers No.1.

Schon, D.A. (1971) *Beyond the Stable State.* W. W. Norton & Co., New York.

Scowcroft, W.R. (1989) Impact of somaclonal variation on plant improvement and IARC/private sector collaborative research In: *Strengthening Collaboration in Biotechnology: International agricultural research and the private sector.* Proceedings of a Conference held 17-21 April, 1988 in Rosslyn, USA. Bureau for Science and Technology, Office of Agriculture, Agency for International Development (AID), Washington, USA, 129-139.

Senez, J.C. (1987) The new biotechnologies: Promise and performance. *The Unesco Courier,* March, 4-12.

Simmonds, N.W. (1986) A short review of farming systems research in the tropics. *Experimental Agriculture,* 22, 1-13.

Stolp, A. and Bunders, J.F.G. (1989) Biotechnology: wedge or bridge. *Trends in Biotechnology,* 7(1), S2-S4.

Stolp, A. and Langeveld, H. (1990) Country case study Zimbabwe. Supplement to Bunders *et al.* (1990). Department of Biology and Society, Vrije Universiteit Amsterdam, The Netherlands.

Stribley, D.P. (1989) Present and future value of mycorrhizal inoculants In: *Microbial Inoculation of Crop Plants*, R. Campbell and R.M. Macdonald (eds). Special publications of the Society for General Microbiology, **25**, IRL Press, Oxford, England, 49-65.

Suarez, R.A. (1984) Towards management of interlinkages In: *The management of interlinkages*, R.A. Suarez (ed.). 12-37.

Tartaglia, J. and Paoletti, E. (1988) Recombinant vaccinia virus vaccines. *Trends in Biotechnology*, **6**, February, 43-46.

Towalski, Z. and Rothman, H. (1986) Enzyme technology In: *The Biotechnological Challenge*, S. Jacobssen, A. Jamison and H. Rothman (eds). Cambridge University Press, 37-76.

UNDP (1989) Plant biotechnology including tissue culture and cell culture. *UNDP Programme Advisory Note*, New York, July.

UNIDO (1984) Design and Evaluation: A manual of policies, procedures and guidelines for UNIDO-executed projects and programmes, Volume 1 - Projects. United Nations Industrial Development Organisation, PC.31/Rev.1, May.

Van Brunt, J. (1987a) Pheromones and neuropeptides for biorational insect control. *Bio/Technology*, **5**, January, 31-36.

Van Brunt, J. (1987b) Bringing biotech to animal health care. *Bio/Technology*, **5**, July, 677-683.

Van Brunt, J. and Klausner, A. (1987) Pushing probes to the market. *Bio/Technology*, **5**, March, 211-221.

Van Rijn, J. (1991) Estudio especial Bolivia. Supplement to Bunders *et al.* (1990).Department of Biology and Society, Vrije Universiteit Amsterdam, The Netherlands.

Von Gizycki, R. (1987) Cooperation between medical researchers and a self-help movement: The case of the German Retinitis Pigmentosa Society In: *The Social Direction of the Public Sciences: Causes and consequences of cooperation between scientists and non-scientific groups*, S. Blume, J. Bunders, L. Leydesdorff and R. Whitley (eds). D. Reidel Publishing Company, 75-88.

Vroemen, B., Enzing, C., Ledeboer, A., Rip, A., and T. van de Sande (1989) Strengths and weaknesses of biotechnology in developing countries: Exploration of methods In: *Report of the Dutch Working Group for the IFIAS Symposium on Biotechnology for Sustainable Development*, 17 May.

Wagner, P. (1987) Social sciences and political projects: Reform coalitions between social scientists and policy-makers in France, Italy and Germany In: *The Social Direction of the Public Sciences: Causes and consequences of cooperation between scientists and non-scientific groups*, S. Blume, J. Bunders, L. Leydesdorff and R. Whitley (eds). D. Reidel Publishing Company, 277-306.

West, D.P., Cramer, J.H., Romero-Severson, J., Ma, Y. and Murray, M. (1989) Application of restriction fragment length polymorphism to plant breeding. In: *Strengthening Collaboration in Biotechnology: International agricultural research and the private sector*. Proceedings of a Conference held 17-21 April, 1988 in Rosslyn, USA. Bureau for Science and Technology, Office of Agriculture, Agency for International Development (AID), Washington, USA, 175-184.

Whitton, B.A. and Roger, P.A. (1989) Use of blue-green algae and Azolla in rice culture In: *Microbial Inoculation of Crop Plants*, R. Campbell and R.M. Macdonald (eds). Special publications of the Society for General Microbiology, **25**, IRL Press, Oxford, England. 89-100.

Wilbanks, T. (1986)   Communications between hard and soft sciences   In: *Interdisciplinary Analysis and Research: Theory and practice of problem-focussed research and development*, D.E. Chubin, A.L. Porter, F.A. Rossini, and T. Connolly (eds). Lomond Publications, Maryland, USA, 131-140.

Wolf, E.C. (1986) Beyond the Green Revolution: New approaches for third world agriculture. *Worldwatch Paper 73*, October.

World Bank (1990) World development report 1990: Poverty. Oxford University Press.

Yuthavong, Y. and Bhumiratana, S. (1989) National programs in biotechnology for Thailand and other Southeast Asia countries: Needs and opportunities In: *Strengthening Collaboration in Biotechnology: International agricultural research and the private sector*. Proceedings of a Conference held 17-21 April, 1988 in Rosslyn, USA. Bureau for Science and Technology, Office of Agriculture, Agency for International Development (AID), Washington, USA, 33-52.

Index